Ordinary Cities

With the urbanisation of the world's population proceeding apace and the equally rapid urbanisation of poverty, urban theory has an urgent challenge to meet if it is to remain relevant to the majority of cities and their populations, most of which are outside the West.

Ordinary Cities establishes a new framework for urban studies, one which considers all cities to be 'ordinary'. It cuts across a longstanding divide in urban scholarship between Western and other kinds of cities, especially those labelled third world, and contests approaches which categorise or rank cities. The book suggests that the two concepts of modernity and development instituted a decade's long division within the field of urban studies between understandings of Western and other cities and that they continue to enable the fiction that cities can be ranked against one another. Key urban scholars and debates, from Georg Simmel, Walter Benjamin and the Chicago School to Global and World Cities theories are explored, together with anthropological and developmentalist accounts of poorer cities. Tracking paths across these previously separate academic literatures and policy debates, the book traces the outlines of a post-colonial approach to cities. It draws on evidence from a range of cities, including Rio, Johannesburg, Lusaka, New York and Kuala Lumpur to ground the theoretical arguments and to provide examples of policy approaches suitable for a world of ordinary cities.

This groundbreaking book argues that if the futures of cities are to be imagined in equitable and creative ways, urban theory needs to overcome its Western bias. The resources for theorising cities need to become at least as cosmopolitan as cities themselves, drawing inspiration from the diverse range of contexts and histories that shape cities everywhere.

Jennifer Robinson is Professor of Urban Geography at the Open University, Milton Keynes, UK.

Questioning Cities

Edited by Gary Bridge, *University of Bristol, UK* and
Sophie Watson, *The Open University, UK*

The 'Questioning Cities' series brings together an unusual mix of urban scholars under the title. Rather than taking a broadly economic approach, planning approach or more socio-cultural approach, it aims to include titles from a multi-disciplinary field of those interested in critical urban analysis. The series thus includes authors who draw on contemporary social, urban and critical theory to explore different aspects of the city. It is not therefore a series made up of books which are largely case studies of different cities and predominantly descriptive. It seeks instead to extend current debates through, in most cases, excellent empirical work and to develop sophisticated understandings of the city from a number of disciplines including geography, sociology, politics, planning, cultural studies, philosophy and literature. The series also aims to be thoroughly international where possible, to be innovative, to surprise, and to challenge received wisdom in urban studies. Overall it will encourage a multi-disciplinary and international dialogue, always bearing in mind that simple description or empirical observation which is not located within a broader theoretical framework would not – for this series at least – be enough.

Ordinary Cities

Between modernity and development

Jennifer Robinson

Routledge
Taylor & Francis Group

LONDON AND NEW YORK

First published 2006
by Routledge
2 Park Square, Milton Park, Abingdon, Oxon OX14 4RN

Simultaneously published in the USA and Canada
by Routledge
270 Madison Ave, New York, NY 10016

Routledge is an imprint of the Taylor & Francis Group

© 2006 Jennifer Robinson

Typeset in Times and Bauhaus by
RefineCatch Limited, Bungay, Suffolk
Printed and bound in Great Britain by
TJ International Ltd, Padstow, Cornwall

British Library Cataloguing in Publication Data
A catalogue record for this book is available from the British Library

Library of Congress Cataloging in Publication Data
Robinson, Jennifer, 1963–
Ordinary cities : between modernity and development / Jennifer
Robinson.
        p.   cm.—(Questioning cities series)
    Includes bibliographical references and index.
    1. Cities and towns.   2. Social history.   I. Title.   II. Series.
HT111.R56   2005
307.76—dc22   2005014089

ISBN10: 0–415–30487–3   ISBN13: 9–78–0–415–30487–0 (hbk)
ISBN10: 0–415–30488–1   ISBN13: 9–78–0–415–30488–7 (pbk)

Contents

Illustrations

Figures

Tables

Preface

I'd like to begin this book in the city I think of as home, Durban, in South Africa. The cover image and the image reproduced below are the work of Tito Zungu who lived much of his life there, although he grew up and spent his old age in a remote settlement a couple of hours outside of the city. He worked as a migrant labourer in a dairy and then as a domestic worker in the neighbourhood in which I grew up. His pictures all feature a sleek aeroplane, often flying over a cityscape. I was half-way through this book when, on a trip back to Durban, one of his drawings caught my eye. It was hanging in a museum better known for its collection of older crafts and art, and I was struck by the excitement the artist so obviously demonstrated for the hyper-modern. Learning a little about Tito Zungu's background only inspired me more. Most of his drawings were on envelopes, as in the image below, which he decorated for migrant workers who, like himself, were keeping in touch with their families and friends in their home places. Later in life, when his work was supported by a local art collector and curator, he drew some larger commissioned pieces, one of which is the cover image of this book. Although his life in the city was spent in modest circumstances, circumscribed by the

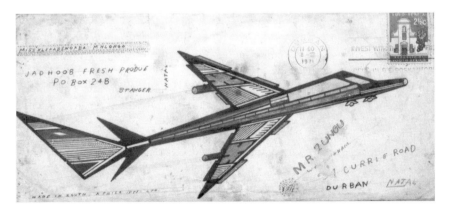

Figure 0.1 Envelope, by Tito Zungu (courtesy of the Killie Campbell Museum, University of KwaZulu-Natal).

racisms of apartheid, Tito Zungu's investment in the modernity of the city – its skyscrapers, speed and energy – was evident in every picture he drew.

Another person who spent a significant part of his life in Durban, Herbert Dhlomo, playwright, poet, journalist and novelist, was also drawn to the excitement of the city. One of a series of poems he penned in Durban, probably during the 1940s, has him declaring, from the Berea Ridge, vantage point for viewing the city:

> I stand and gaze and feel – and marvel! Is
> This then the great city that has planted
> Despair in me? What contrasts jolt in this
> Strange Hive: souls kind and hard; pure Good; great Sins!
> This Hope or Mockery, Lord? Or Joy or Pain?
> For here beneath my eyes lie wonder scenes
> That should ring Joy but only fling me Pain!
>
> (Dhlomo 1985: 364)

Dhlomo was an insightful commentator on South African affairs, working on an independent African newspaper in Johannesburg, and for many years he was the editor of the most significant Zulu language newspaper in Durban. Here, as in some other poems and some of his journalism, he reflects on the deep complexity of South African urbanism. The modernity of the city, as compelling for him as for Tito Zungu thirty years later, was as much a source of pain as joy. The excitement he notes at the end of the poem extracted above, as he views the flashes of the electric lights of the city in the night sky – '... I look / Where Nature's work and Man's mingle or fight – / Up sprout Man's flowers! Electric lights! 'Tis night!' – stands in painful contrast to the sense of exclusion he anticipates as the city's sins of hatred and racism come to his mind.

These two artists from Durban remind us that the excitement and wonder of cities is available to everybody who lives there. And yet as both their lives also remind us, many people are made to feel excluded from these aspects of city life. Racially divided South African cities made modernity a deeply ambivalent experience for many who lived there. But it is the argument of this book that urban theory has also excluded many cities and their citizens from their accounts of the excitement and potential of city life. Theories of modernity, just like South African urbanism, have often reserved experiences of dynamism and innovation for a privileged few, and especially for those wealthy cities and their citizens who have laid claim to originating modernity. In the process, poorer cities and marginal citizens have been profoundly excluded from the theoretical imaginary of urban modernity. Theories of urban modernity, then, have drawn a stark line between 'modern' cities and other kinds of cities, variously described as Third World, perhaps African, perhaps developing/underdeveloped, perhaps colonial. At best these 'other' cities have been thought to borrow their modernity from wealthier contexts, presenting pure imitations rather than offering sites for inventiveness and innovation. This book is a contribution to the task of ensuring that as scholars of cities we end our complicity in such dispossessions. Instead, I will be making the case for an account of urban modernity that tracks its influences and origins far and wide. In fact, I would like to reverse the act of dispossession and refuse the West's claim to ownership of modernity.

In my move from Durban to London, I became aware that the dispossessions of a theory of modernity were being reinforced by some later trends in urban studies. It was then that I realised that the rest of the world didn't necessarily think of Durban, or any other South African city for that matter, in the way I did. Of course, anyone from outside the city was not going to understand how that place could be the centre of my world, a core reference point for life choices and ambitions. But in intellectual terms, the difference in approach was more significant than this: South African cities fitted uneasily into the architecture of knowledge about cities. Scholars of cities there had been drawn to explain the distinctive social relationships and spatial forms that emerged in this racially divided context. To do this we followed literatures on cities in many different parts of the world to appreciate the politics of economic privilege, political power and determined opposition to racism in South African cities. By the time I moved to London in 1995, we were also starting to think about the demands for development that followed on from liberation from apartheid. And yet, for my students at the London School of Economics (LSE), understanding cities in the 'Third World' was all about development. Stories about the modernity of poor cities, their diverse cultural practices and complicated political struggles were somehow seen as betraying the need to do something about the terrible circumstances in which many city dwellers lived.

I was struck, then, by the profoundly different approach adopted by most Western scholars studying cities in poorer contexts, so different from the way in which we had thought of South African cities as part of the world of cities – exceptional, perhaps, in their racism but certainly to be thought of in relation to cities everywhere. By contrast, studies of Third World cities seemed to occupy a distinctive and separate sphere of intellectual enquiry from studies of cities elsewhere and were generally ignored by writers on wealthier cities. Developmentalism, it seemed to me, had compartmentalised the field of urban studies; it had kept scholars from sharing their understandings of cities, from learning from one another.

This book is an attempt to move across the divides in urban studies that have been generated by theories of modernity and by the conceptual apparatus of developmentalism. Along with citizens and artists like Tito Zungu and Herbert Dhlomo, and with cities like Durban in mind, I want to insist that theories of urban modernity belong to all cities and their citizens. And in contrast to the intellectual divides generated by developmentalism, I want to propose a post-colonial urban studies in which scholars of wealthy, Western cities learn about their cities by thinking with scholars and artists from other places; and cities and their citizens everywhere are thought of as participants in the creativity and dynamism of contemporary urban life.

Acknowledgements

Parts of earlier versions of Chapter 1 have been published in: 'Cities between Modernity and Development', *South African Geographical Journal*, 86 (2004) 1: 17–22; and 'In the Tracks of Comparative Urbanism: Difference, Urban Modernity and the Primitive', *Urban Geography*, 25 (2004) 8: 709–23.

Earlier versions of parts of Chapters 4 and 5 appeared in 'Global and World Cities: A View from Off the Map', *International Journal of Urban and Regional Research* 26 (2002) 3: 531–54; and 'Urban Geography: World Cities, or a World of Cities', *Progress in Human Geography* 29 (2005) (in press).

Many thanks to the following people for comments and advice on aspects of this work at various stages: Gareth Jones, Mona Domosh, Tim Bunnell, Abdoumaliq Simone, Luciana Martins, Mauricio Abreu, Edgar Pieterse, John Allen, Garth Myers, Gilles Duranton, and three anonymous reviewers who read the entire manuscript. My research in Johannesburg was greatly assisted by Rashid Seedat and I also owe a debt of thanks to the many other busy officials and councillors who made time to talk to me about their experiences. Teresa Dirsuweit and John Spiropoulos shared time in Johannesburg and taught me to understand their city a little better and to enjoy it a lot. Sue Parnell and Barbara Lipietz have both been working on similar issues in Johannesburg and I have benefited enormously from reading their work and talking with them about the details of Johannesburg's development politics. Yvonne Winters and Nellie Somers at the Killie Campbell Library of the University of KwaZulu-Natal helped with finding the cover image. Andrew Mould and his assistants Anna Somerville and Zoe Kruze at Routledge have played an important role in encouraging me first to prepare the proposal and then to get the book done in good time. Jan Smith, Rebecca White, Michele Marsh, Annie Williamson-Noble and Neeru Thakrar, all at the Open University, have provided excellent (and always good-humoured!) administrative support for all the different aspects of the projects that led to this book. I owe thanks to the Geography Discipline and the Faculty of Social Sciences at the Open University which supported me with time and money for research and writing; the spirited intellectual environment there has certainly challenged me to do better. Finally, Steve Pile read and commented on the whole draft and has been a constant intellectual companion during the time I have been working on this book, improving the arguments, seeing me through its ups and downs and cheering me up by taking me to watch lots of vampire and alien movies. Thank you!

Introduction
Post-colonialising urban studies

It is the argument of this book that all cities are best understood as 'ordinary'. Rather than categorising and labelling cities as, for example, Western, Third World, developed, developing, world or global, I propose that we think about a world of ordinary cities, which are all dynamic and diverse, if conflicted, arenas for social and economic life. Whereas categorising cities tends to ascribe prominence to only certain cities and to certain features of cities, an ordinary-city approach takes the world of cities as its starting point and attends to the diversity and complexity of all cities. And instead of seeing only some cities as the originators of urbanism, in a world of ordinary cities, ways of being urban and ways of making new kinds of urban futures are diverse and are the product of the inventiveness of people in cities everywhere.

We will see that there are some serious effects consequent upon labelling cities, placing them in hierarchies or dividing them up according to levels of development. This book will outline some of these effects and also explain why it is that the study of cities arrived at a place where categorising, ranking and labelling cities has become so prominent and so damaging to the future prospects of cities everywhere. Understanding cities as ordinary, it will be argued, opens up new opportunities for creatively imagining the distinctive futures of all cities.

Ordinary Cities forms the basis for a new, post-colonial framework for thinking about cities, one that cuts across the long-standing divide in urban scholarship between accounts of 'Western' and other kinds of cities, especially cities that have been labelled as 'Third World'. It is in the spirit of a post-colonial critique that this book promotes the case for 'ordinary cities' (Amin and Graham 1997), for an urban theory that draws inspiration from the complexity and diversity of city life, and from urban experiences and urban scholarship across a wide range of different kinds of cities. This is in strong contrast to much urban theory, which has taken its primary inspiration from cities in the West and which has tended to privilege certain experiences of these places.

Extending a post-colonial critique of social theory to the study of cities, this book sets an agenda for a new generation of urban scholarship that will move beyond divisive categories (such as Western, Third World, African, South American, South-East Asian, or post-socialist cities) and hierarchies (such as global, alpha or world cities). I hope to establish the basis for a post-colonial urban theory that will challenge the colonial and neo-imperial power relations that remain deeply embedded in the assumptions and practices of contemporary urban theory. These are certainly evident in the practice of dividing, categorising and assuming hierarchical relations amongst cities, but they are also visible in accounts of urban modernity – the creativity, dynamism and innovativeness of cities – which have assumed a privileged relationship with certain wealthy, Western cities. By contrast, a post-colonial urban studies would draw its inspiration from all cities, and all cities would be understood as autonomous and creative.

With the urbanisation of the world's population proceeding apace and the equally rapid urbanisation of poverty, urban theory has an urgent challenge to meet if it is to remain relevant to the majority of cities and their populations, most of which are now outside the West. The existing bias in urban studies towards Western cities and the relegation of cities in poor countries to residual categories (or, in fact, completely off the map of some approaches to cities) makes the irrelevance of urban theory a real possibility in the light of global trends in urbanisation. A post-colonial revisioning of how cities are understood and their futures imagined is long overdue. But to reconfigure the field of urban studies, scholars need to be prepared to do the hard work of examining some of the basic assumptions and key concepts that somewhat surreptitiously (but also most obviously, when you start looking) divide and limit the field of urban studies. In seeking to do this, the book will address both the cultures of cities, specifically the close association between cities and cultures of modernity, and the economies of cities as these have been framed through the ambitions of city managers, residents and policy-makers to promote urban development.

It is these two axes of debate in urban theory that have been important in dividing the field of urban studies between Western and Other cities: celebrations of urban 'modernity' and the promotion of urban development. Together these have produced a deep division within urban studies between those cities that have been seen as sites for the production of urban theory and those that have been portrayed as objects for developmentalist intervention. These latter cities have provided the grounds for demarcating difference in a system of hierarchical relations amongst cities. Perhaps most importantly, together these conceptual fields continue to ascribe innovation and dynamism – modernity – to cities in rich countries, while imposing a regulating catch-up fiction of modernisation on the poorest. This book will consider both the concepts of modernity and development in some detail, tracking paths across previously separate academic literatures and policy debates in order to seek out more cosmopolitan resources for building a post-colonial understanding of cities.

There is an important opening right now, as the rise of post-colonial approaches to social theory in the Western academy valorises the production of knowledge in other places and encourages a rethinking of many fundamental concepts of Western scholarship. It is, of course, a loss to Western theory to have ignored for so long the vibrant scholarship in and on places beyond its usual realm of operation. But in a connected-up world this lack of engagement is a loss to all of us. Following post-colonial critics such as Dipesh Chakrabarty, Gayatri Chakravorty Spivak and Dilip Goankar, the task is not so easy as to simply abandon the concepts of Western academia, which have already profoundly hegemonised the field of intellectual production. Rather, it is to engage critically and to insist on the dual move of at once parochialising Western knowledge, excavating its local and limited origins, but also and perhaps more importantly provoking a more cosmopolitan engagement with experiences and scholarship elsewhere. By this I mean to build on James Clifford's (1997) idea of 'discrepant cosmopolitanisms', rather than a universalising or homogenising cosmopolitan impulse (see Cheah and Robbins 1998). Clifford's interest is in de-localising professional anthropology, partly by insisting that cultures are not (only) localised and that the village-based ethnography is a limited and misleading research tool (see also Gupta and Ferguson 1999). Connections and travels beyond the local are long standing and constitutive of local cultures all over the world: an important consideration for any urban scholar (Smith 2001). But, importantly for this book, he also wants to encourage anthropologists to consider the trajectories of their own practices and analyses. Urbanists, too, could find it valuable to think about the contrast between the restricted spatialities of their theories – the geography of urban theory – and the diverse cosmopolitanisms of the cities they write about.

In the South African novel, *Welcome to Our Hillbrow* (an inner-city area in Johannesburg), Phaswane Mpe (2001) portrays 'our' Hillbrow as belonging to many different people from all over the place – a composite of many different migration paths, connections to other parts of the city, the countryside, the continent. It is the argument of this book that we need a form of theorising that can be as cosmopolitan as the cities we try to describe. This would be a form of urban theory that can follow the creative paths of urban dwellers – across the city or around the world – as they remake cities (Simone 1998) and that can draw on the transformative potential of shared lives in diverse, contested – ordinary – cities to imagine new urban futures. Anything less will mean a failure on the part of urban theory to contribute to the imagination of better futures for cities everywhere.

BETWEEN MODERNITY AND DEVELOPMENT

In this book I will be exploring two concepts, modernity and development, that I think play a key role in perpetuating the division of urban studies I have outlined. They have been central to the analysis of city life, and of city futures for some time – although for different periods of time – and each draws the world of cities into a range of categorisations and differentiations that, I argue, limit the potential of contemporary urban theory. By urban modernity I mean the cultural experience of contemporary city life and the associated cultural valorisation and celebration of innovation and novelty. And by development I mean the ambition to improve life in cities, especially for the poorest, along certain policy-informed paths. A political investment in development, and the institutional promotion of development as a way of improving life in poor countries, following Escobar (1995), we can call, 'developmentalism'.

These two concepts are closely entwined. Together they work to limit both cultural imaginations of city life and the practices of city planning. For without a strong sense of the creativity of cities, of their 'modernity', the potential for imagining city futures is truncated. And, as much of the production of knowledge and insight about cities in poorer places has been tied to their poverty and concerned with the things they lack, imagining creative, distinctive future trajectories for these places has not been easy. The concept of modernity, then, profoundly reinforces the work of developmentalism in urban studies, as the 'modern' has been theoretically aligned primarily with Western cities, or with the export and proliferation of a supposedly 'Western' modernity around the world.

Both modernity and developmentalism are products of a colonial past. Current meanings of the term 'modern' have been largely defined by early twentieth-century Western scholars. To be crude, modernity could be understood as simply the West's self-characterisation of itself in opposition to 'others' and 'elsewheres' that are imagined to be not modern, an opposition that was strongly reinforced through the mundane practices of colonisation (King 2004: 71). Such a colonial imagination was present at the birth of urban studies. As key Western writers on cities such as Georg Simmel and Robert Park sketched the theoretical foundations for urban studies at the turn of the twentieth century, they did so in the tracks of colonial practices of racialisation and cultural difference. On this basis they erected a fantasy about the cities they knew as being creative, dynamic, modern places. This fantasy about the nature of urban experiences in the West persists and, as we will see in the Chapter 1, it does so on the basis of designating other places and other people – other cities – as 'not-modern'.

In contemporary Western urban theory there is a deep historical borrowing from literatures of the first decades of the twentieth century, especially concerning understandings of urban modernity. Colonial prejudices from this era have been sedimented into contemporary theory. Most surprisingly, this has taken place despite the fact that there were lengthy debates contesting these assumptions during the middle decades of the twentieth century. This scholarship, mostly from the 1940s through to the late 1960s, although still framed by Western accounts of urban modernity, was encouraged by the comparativism of anthropology and found license to critique, disorient and displace the influential accounts of urban modernity that had emerged from

Western Europe and North America. A generation of urban scholars embarked on a massive outpouring of work in the field of comparative urbanism and contested both the ethnocentricism of Western scholarship and disproved many of the racist assumptions of dominant approaches to cities.

But these debates and the scholars who contributed to them have been largely forgotten in the canons of contemporary Western urban theory. The contemporary fin-de-siècle canon of urban studies leaps across this period of scholarship as if it were a barren canyon of intellectual endeavour, reaching back to the previous turn of the century, to writers from Berlin and Paris, stretching lithely over to Chicago and then blanking out on the journey towards the present. The agent of forgetfulness in the landscape of urban theory, I suggest, is the damaging divide implemented by developmentalism. For the past few decades urban studies has fixated on the categories of success (wealthy global cities) and the categories of a noir futuristic urban genre of decline and despair (the poor mega-cities). In doing this it has divided itself and built deep into its intellectual reason the differences of a world in development.

The inheritance of developmentalism within contemporary urban studies, then, is equally as strong as that of modernity. With long roots in colonial paternalism and a rise to prominence in the context of decolonisation and powerful neo-imperial ambitions on the part of wealthy nations, developmentalism has functioned to make the experiences of cities in developed and developing, or underdeveloped, contexts appear broadly incommensurable (Cowen and Shenton 1996). The project of development has focused attention on the differences between cities that are assumed to be at different stages of advancement and has embedded hierarchical assumptions about the relations amongst cities into the analysis of cities at an international scale. Ironically, this was as much an outcome of a radical and progressive approach to cities most associated with scholars in South America – a Marxist critique of underdevelopment – as it was the result of the powerful effects of Western development policies. The field of debate on cities and their futures is now sundered between categories of cities whose trajectories and experiences are considered unrelated to each other. Many assumptions routinely introduced into urban writing urgently need revisiting, and for this to happen cities that have been kept apart by urban theory need to be brought within the same field of critical inquiry.

I set out in this book to contribute to what will be, I think, a long project to refute these divisions within the field of the urban.[1] Here I aim to mark out a few exploratory trails across the divides of modernity–tradition and development–underdevelopment and to ride more confidently along roads well travelled until the 1960s, in the footsteps of writers on the city who have never been canonised. Remembering forgotten work on a diversity of urban contexts is an important strategy in generating a post-colonial urban studies. But it will also be important to find ways to analyse cities that deal differently with their differences – gathering difference as diversity rather than as hierarchical division.

My intention is to proliferate opportunities for urban studies to assess the horizon of meaning within which it is currently operating. Most aggressively, and much inspired by Henri Lefebvre, I want to 'detonate' the assumed meanings of 'modernity' and 'development'. I also want to insist that theorising about cities should be more cosmopolitan, should be resourced by a greater diversity of urban experiences. And I want to achieve a collective refusal of the categories and hierarchising assumptions that have left poor cities playing a punitive game of catch-up in an increasingly hostile international, economic and political environment.

Responding to this predicament, this book will establish how located and parochial assumptions about the nature of urban modernity have been universalised and will consider how the term might be reinvigorated for a more cosmopolitan urban theory. It will also demonstrate how developmentalism has divided a field once geographically eclectic in its comparativism. And, most importantly, I hope that the analysis that follows will help to revitalise how scholars and practitioners think about improving city life and city economies, drawing on insights and experiences from a more extensive range of urban contexts. Consequently, I envisage an urban theory substantially committed to comparative work, suspicious and cautious about deploying categories and hierarchies and eager to promote strategies for city improvements that build on their distinctive and individual creativities and resources.

DISLOCATING MODERNITY, DIVERSIFYING DEVELOPMENT

Because the current predicament of a divided urban studies can be traced back to the twin concepts of modernity and development, this book traverses literatures that are seldom brought together. Not just because the juxtaposition of concerns with poorer and wealthier cities is relatively unusual in the literature, but also because I am suggesting that accounts of the cultural politics of urban space matter very much for proposals for city development. Accounts of economic clustering and flânerie don't often find comfortable alignment, but I am convinced that without the imaginative resources of autonomous vitality that a revised conceptualisation of the modernity of cities – all cities – could offer, urban studies will continue to be complicit in limiting the potential for improvement and growth in cities in poorer countries and the prospects of poor people in cities everywhere. Rather than assuming a developmentalist perspective on the challenges of life in the poorest cities, writers such as Karen Hansen (1997) and Abdoumaliq Simone (2004) have argued for an appreciation of the close entwining of cultural practices and economic challenges. This brings into view how distinctive ways of urban life in different cities both enable diverse livelihood strategies and creatively remake senses of self. And from the position of policy-makers and urban managers, assuming that cities are vital and dynamic sites where

citizens are shaping autonomous and inventive futures would offer considerably more scope for creative and relevant interventions than copy-cat policies
that aim to reproduce the experiences of cities elsewhere. The book moves,
then, from a reformulation of the concept of modernity as applicable to all
cities, to a reframing of urban development as a series of challenges that face
the wealthiest and the poorest cities. In both cases, a strong argument will be
made for creative learning across the experiences of diverse kinds of cities.

The first few chapters of the book undertake the important ground-
clearing exercise of dislocating and reconfiguring the concept of modernity
for use in the twenty-first century. I am not satisfied with accounts of urban
cultures that announce the presence of 'alternative' modernities (Goankar
2001) or that pluralise the experience of the modern. For all of these accounts
maintain as primary the 'modern' invented in the West. It is in reference to
the inventions of the West that transfigurations, adaptions and hybridisations
of the 'modern' are observed. I want to decouple understandings of the
modern from its association with the West, and I want to dislocate accounts
of 'urban modernity' from those few big cities where astute observers
elaborated on the broader concept of 'modernity', placing it in a privileged
relationship to certain forms of life in these places.

I develop a concept of the modern that sees many different cultures in
many different places as enchanted by the production and circulation of
novelty, innovation and new fashions. These cultures, some of which have had
the awful misfortune to be characterised as 'primitive' by generations of
Western scholars, share with many people in the present and in a diversity
of contexts, an enjoyment of and attachment to new things – or things that
are new to them. This is not the only reaction that people and cultures have to
innovation. Novelty can be dangerous, disruptive and is frequently dis-
avowed. There is certainly a politics to modernity. An investment in things
that have worked well, or practices that have served certain interests for some
generations (but not forever – timelessness is not an option in any society)
are very good reasons to desist from engaging in new things. But opposing
'tradition' to 'modernity' is the first and largest error of existing accounts of
the modern. And viewing the embrace of novelty as 'innovative' in Western
contexts but 'imitative' in others is its second profound error. Together these
errors have produced an account of urbanism and urban modernity that has
travelled very poorly.

Chapter 1 excavates how the Chicago School theorists established an
understanding of urban experience through a strong contrast with tradition.
The account of urban modernity that emerged from their work, which is
the workhorse of current urban theory, postulates a modern 'here and now'
against a traditional 'there and then'. The past and a haphazard range of
other places fulfilled the function of making some urbanites feel very
modern. These fantasies enhanced urbanists' sense of elation about what
they liked to think of as the novelty of their experiences. But the parochial
nature of this fantasy and its dependence on seeing other places and periods

as backward and slow, is a most inadequate foundation for a post-colonial urban studies. Inspired by Walter Benjamin's rich dialectical imagination, the rest of the chapter considers whether there is scope for rescuing the concept of modernity from the ruins of urban theory. In this task, we are much aided by the work of anthropologists who took the Chicago School hypotheses to cities around the world, and found them limiting.

The selective memory of the urban canon, though, has forgotten this series of debates about the nature of 'urbanism' around the world. Scholars conducted detailed studies on the complexities of ways of being 'modern' and 'urban' in many different cities around the world, critiquing and extending the analyses from Europe and America. And all of these dispel the easy binary of modernity and tradition. Drawing on their work, and following Benjamin for a moment, we could imagine that we had observed a 'dialectics at a standstill' in the interaction between the Western theories of modern urbanism, with their fantastic representations of traditional or folk cultures, and the range of emerging vibrant, dynamic urban cultures in cities as represented by urban anthropologists. These quite different urban imaginaries came together in writing about cities around the world in the outpouring of comparative urbanism from the 1940s through to the late 1960s and provoked a reconfiguration of ways of understanding the relationship between urbanism, modernity and tradition. Had urban studies not been divided by developmentalism, these accounts might have been drawn on by more recent writers to offer insight into the possibilities for different kinds of urbanism and diverse urban ways of life. Cities everywhere might have been imagined and lived differently, perhaps as more sociable, more variable, less prone to producing anomie, less dependent on private initiative if, for example, African urban experiences had been allowed to inform urban theory. Can we find in these debates the resources to begin a reconfiguration of the concept of urban modernity, fitting for a post-colonial urban studies?

Chapter 2 sets out to do this and explores some specific aspects of debates provoked by this mid-twentieth-century comparative urbanism. The chapter focuses on one of the analyses of urban life frequently developed on the basis of the works of Georg Simmel, Robert Park and Louis Wirth: that to survive in the modern city, with its many shocks and challenges, with the density of personal interactions and the ever-tumultuous presence of newness, urbanites must become characteristically blasé, or indifferent. A large number of scholars, both within and outside the West, contested this rather simplistic claim. Yet it remains a constant component of contemporary urban theory and is often reproduced in urban readers, while the critiques of comparative urbanists have largely been ignored. This chapter establishes the importance of redrafting the history of urban theory, drawing on these rich resources of comparative research.

Comparative urban studies may have been shipwrecked on the reef of developmentalism some time in the late 1960s, but this chapter suggests that one strategy for instigating a post-colonial urban studies would be to build on the insights of these now-neglected writings. Tracking a direct critique of the Chicago School by a group of Africanist anthropologists (the 'Manchester School') the chapter suggests that urban dwellers, rather than being blasé, creatively engage with the difficulties and openings that cities present in a number of different ways. In a wide range of contexts, but specifically drawing here on work on the Copperbelt of Zambia in central Africa, we learn that urbanites have generated fictive kin, eagerly sought to make connections where none really existed, carefully nurtured neighbours and family, built communities and defended difference. Cosmopolitan theories of urbanism will reflect this diversity of urban experience. The chapter turns to reflect this critique back onto early twentieth-century urban theory, this time drawing on Walter Benjamin and Georg Simmel to excavate the sociable interactions and emotional interdependencies of life in northern European cities, including in that apparently most anonymous of urban phenomenon, the city crowd.

In the light of these critiques, Chapter 3 works with a reinvigorated concept of urban modernity, one that distributes innovation and creativity promiscuously across diverse kinds of cities. Moving on from comparative methodologies, this chapter explores the prolific circulation of the cultural practices and artefacts of urban modernity around the globe. Much of the subject matter concerns the proliferation of modernist and internationalist urban design across the world's cities. In some cities, early twentieth-century modern architecture added up to evidence of authentic forms of urban modernity: powerful Western cities drew on inspiration from many different parts of the world to create new kinds of built environments and, in some cases, produced distinctive skylines that became iconic of urban modernity. In other places, though, creative experiments with modernism and other elements of urban design have been assigned simply to mimicry.

Where Chapter 2 tracked paths across the modernity–tradition divide that has framed the field of urban studies, drawing on comparative urbanism, Chapter 3 engages with the circulation of modernist urban design and development through different urban contexts. As a tactic towards a post-colonial urban studies, tracking the circulation of urbanisms brings different cities within the same theoretical field. But while the modernity–tradition distinction marked certain cities and urban cultures as outside the realm of the modern (and the urban), the proliferation of modernism across the sky-lines of cities everywhere apparently marks certain cities as purely imitative, only mimicking the innovations of other cities, other cultures. Why is internationalism in New York evidence of that city's vitality and creativity, yet Rio de Janeiro's dynamic architectural heritage always carries the spectre of Europe? This chapter once again tries to track a different path across a version of difference within the world of cities, gathering difference as diversity, rather than as hierarchical categorisation or incommensurability. The travels of modernist design will be traced with a view to dislocating its associations with a small range of Western cities and to dispossessing the West of its claims to be the originator of both modernism, and modernity.

Architectural modernism and internationalism signify the existence of a global circuit of urban production, design and imagination. Into the late twentieth century, the imagination of the globe as increasingly connected up through communications, trade and organisations (firms as well as international institutions) has come to dominate urban studies. Yet, as Chapter 4 suggests, the 'global' imagination of the world's cities in the last few decades of the twentieth century rests once again on a developmentalist categorisation of differences amongst cities. Quite ironically, even as transnationalism has become the characteristic feature of late-twentieth-century urban experience, cities around the world have once again been hierarchically ordered and categorically divided by urban theorists, with harmful effects for rich and poor cities alike. Chapter 4 engages with the powerful global and world cities hypotheses and argues that in the interests of a post-colonial and politically progressive urban theory, it is long past time to abandon these labels and their associated categories. I suggest the emerging interest in 'ordinary cities' as a far less destructive way of conceptualising the field of the urban. Bringing together the analysis so far in the book, ordinary cities emerge here as diverse, differentiated and contested, shaped by processes stretching far beyond their physical extent, but also by the complex dynamics of the city itself. This chapter suggests that it is the ordinary city that is brought into view by a post-colonial critique of urban studies.

Whereas global and world cities approaches focus on small elements of cities that are connected into specific kinds of economic networks, and developmentalist approaches tend to emphasise the poorest, least well-provisioned parts of the city, the ordinary city approach brings the city 'as a whole' back in to view or, more properly, the city in all its diversity and complexity. This implies a stronger reterritorialisation of the imagination of urban studies around the individual city, or city-region, rather than its immersion in recounting transnational flows. At this urban scale, the challenges of addressing the diversity of city economies and societies and of responding to the quite divergent needs gathered together in any city are readily apparent. The narrow economic reductionism of much global- and world-cities literature can be directly redressed by bringing the city back into urban theory.

On this basis, Chapter 5 draws our attention to spatialities of cities that have been occluded in recent studies. Whereas a lot of emphasis in the global- and world-cities literature has been on flows and networks between cities, or on the emergence of localised concentrations of economic activity, or the fragmentation of social life in cities, this chapter aims to bring the city back in to view, in all its complexity and diversity. Cities are sites for political contestation; they are platforms for economic activity; they are arenas for political redistribution. There is little chance that we could even start to capture anything like the 'city as a whole', given the intense complexity of any city, and the dense webs of interconnections that make up any urban experience. But taking a city-wide view brings back into question issues that have been abandoned by mainstream urban theory, even as they have become crucial elements of policy-making and urban development. Learning from the field of urban development studies and international urban development policy, the importance of thinking across the diversity and complexity of cities is established.

The consequences of the ordinary cities approach for understanding urban policy are significant and form the focus of the final chapter. The ambition is to explore different tactics for promoting urban development. These would be tactics that release poor cities from the imaginative straightjacket of imitative urbanism and the regulating fiction of catching up to wealthier, Western cities that categorising and hierarchical approaches to cities produce. Instead, a more cosmopolitan and transnational approach to improving city life is proposed and is emerging in the practices of urban managers. Outside of a hierarchical and categorising understanding of urban experiences, urban managers and citizens can draw on the distinctiveness of their own city, and on the imaginative resources of city life across the world to inspire their attempts at creative interventions. As one possible methodology for a post-colonial urban studies, this chapter will track across the experiences of wealthier and poorer cities, to illustrate the potential for new geographical alignments in the formulation and analysis of urban policy.

In this regard, a post-colonial urban theory challenges Western-centric accounts of how and why innovation and growth might occur in cities. These accounts have come to stress specialisation and sectoral clustering as the basis for creativity and innovation in cities – accounts primarily based on Western experience that are usually pessimistic about growth prospects for most cities around the world. A more cosmopolitan approach to understanding cities can encourage a different account of the resources for and characteristics of urban development. In Chapter 6 the focus will be on the opportunities for interaction and connection across the diverse economic activities that charac-terise ordinary cities. Instead of or, more properly, in addition to localised clusters of specialised economic activity, the economies of ordinary cities will be seen to depend to a large extent on the most general urban agglomeration economies. And the broader urban context, including questions of poverty, inequality and cohesion, emerges as deeply entwined with the economic futures of cities. As this chapter explains, this offers scope for interventions in support of economic growth that also have the potential for redistributive outcomes. Linking local and transnational economies; tying together informal and formal sectors; realising and building on the creativity and dynamism of the diversity of city economies: these will be the key themes of this chapter.

The Conclusion to the book rehearses some of the tactics that urban scholars might draw on as we seek to postcolonialise our theories and our practices. Here I will consider how the postcolonialising strategies developed in this book can be put to work in the ongoing research and thinking associated with urban studies. I will also draw together the implications of the book for thinking about the geographies of the city. An account of the diverse spatialities of cities weaves through the book, informing the critique of both modernity and development and establishing the grounds for new imaginations of urban development. Finally, the Conclusion will reflect on the real difficulties of finding a path between universalism and incommensurability; dealing differently with the differences amongst cities is the core challenge for a post-colonial urban studies. Assuming that all cities are ordinary is only a starting point for producing new ways of thinking about the world of cities.

1 Dislocating modernity

INTRODUCTION

Perhaps paradoxically, at the heart of many of the foundational texts of Western urban theory we find the figure of the primitive. Identifying the dynamism and modernity of the Western city has often depended on figuring some other places and people as traditional, at best 'rural', at worst 'primitive'. Contemporary thinking about cities silently reproduces this idea of an external 'savagery' that sustains the fantasy of (Western) urban modernity. It does more, though. It has come to support a hierarchical analysis of cities in which some get to be creative, and others deficient, still tainted by the not-modern, placed on the side of the primitive. This chapter teases apart the westernness of accounts of the modernity of city life. Western urban theories have depended upon a profoundly parochial understanding of what it means to be modern in order to frame their apparently universal understandings of urban experience. By contrast this chapter sets in train the critical resources for an account of cities that can escape – or at least start to distance itself – from the heritage of ethnocentric assumptions in urban theory that have bothered urban anthropologists for so long (Southall 1973, King 1990). The critical tactic here is to dislocate accounts of urban modernity from the West and to encourage understandings of all cities as potential sites of creativity and innovation.

In order to describe a particular time as modern, ideas about a new historical era or the emergence of a strong cultural investment in novelty and innovation have been closely tied to a sense of historical progression and progress. Accounts of modernity have commonly described the modern era, or modern people, as having a sense of historical time, based on new, rational techniques for the ordering of time and space, and as drawing on a rationalist understanding of events to inform inventiveness and progress. This is contrasted strongly with previous eras (and other peoples), which are portrayed as having a mythical sense of time and a mystical and religious sense of causality with dominant and static traditions unsuitable for informing social and technical progress. In a deeply colonial move, it is the West that has been seen as the site of modernity and other places that have been entrained as not-modern or less modern through the transformation of historical time into geographical difference (McClintock 1995).

Modernity is, then, a deeply problematic concept for the project of forging a post-colonial urban studies. It allies the emergence of certain historically specific social formations with the idea of progress; and it aligns this sense of progress with certain places. In setting out to post-colonialise urban studies, we might be tempted to assume that the concept of modernity has deeply flawed origins; we might consider that the best move would be to delete it from our repertoire of interpretations (Thrift 2000). But, on the other hand, it has such a resonance with many different cultural projects around the world that this seems actively unhelpful. And as some commentators have argued, 'to announce the general end of modernity even as an epoch, much less as an attitude or an ethos, seems premature, if not patently ethnocentric, at a time when non-Western people everywhere begin to engage critically their own hybrid modernities' (Goankar 2001: 14).

More than this, though, it is the argument of this chapter that theorising cities has been so deeply entwined with Western understandings of modernity that if we are to post-colonialise urban studies we have little choice but to attend to the concept, even if it is to dismantle and dislodge it. Simply wishing it away is quite likely to leave intact unstated but profound assumptions within urban theory that exclude some cities from the achievements of modernity. While the concept of modernity places some cities at the top of a hierarchy of creativity and dynamism, it consign others to a future of mimicry or backwardness.

Following Timothy Mitchell (2000), we might therefore consider the concept of modernity to be a 'nondisposable fiction' (2000: 12), one which we might not like, but that we can't live without. For centuries, the concept of modernity has framed Western accounts of contemporaneity and change. And as Peter Osborne notes,

> Born like capitalism out of colonialism and the world market, 'modernity' as a structure of historical consciousness pre-dates the development of capitalism proper [. . .] As our primary secular category of historical totalisation it is hard to see how we can do without it in one form or another [. . .] We have no option but to rethink 'modernity' as the transformation of the conditions of its existence gathers pace with time
>
> (1992: 40)

Seeking to transform the concept of modernity for post-colonial times, many writers suggest that we could pluralise it to produce a notion of hybrid modernities, different in different places, or 'doubled' through a difference within (Gilroy, 1993, Goankar 2001). My preference is to turn our sights on understanding how the idea of modernity came to have a privileged association with particular places and times. In terms of this book, that means we will need to consider how the idea of modernity came to be tied both to a specific context and to a certain historical phenomenon: Western urbanism. I want to see if, following Osborne, we can refigure our conceptualisation of modernity beyond the historically specific processes that support the profoundly parochial identification of what it means to be modern that has emerged within Western urban theory. I make no claims to actually arriving at a novel, post-colonial or cosmopolitan account of urban modernity in this chapter, although Chapter 3 offers a stronger set of arguments to this end. But I hope to indicate that a post-colonial revision (rather than simple refusal) of the concept of urban modernity is a generally plausible proposition and to suggest some lines of enquiry towards this aim.

This chapter therefore starts off by considering the claim that current theoretical understandings of modernity, and urban modernity specifically, are parochial (Western) conceptualisations parading in the guise of a universal account. As a prime example of ethnocentric theorising, urban studies has placed the assumed characteristics of the Western city – as dynamic, individualising, rational – in strong contrast to 'tradition', which was portrayed as static, communal, in thrall to the sacred, and either outside the West, or definitively in its past.[1] In search of some paths beyond this static categorisation of some cities as modern and dynamic, in contrast to the primitive or tradition, I turn to the work of Walter Benjamin who offers us an alternative, more dialectical vision of the relation between modernity and its postulated others, tradition and primitivity. Read in this light, a field of comparative urban studies, dominant through the 1940s to 1960s, takes on a new meaning. As the 'traditional' subjects of anthropology moved to town and anthropologists followed them to conduct research on cities around the world, any easy dichotomies between the urban and traditional were disturbed. Restaging, in a dialectical fashion, the encounter between the Chicago School and mid-twentieth-century urban anthropology, we can start, in the following chapter, to recover from the ruins of urban theory the potential for a different configuration of urban modernities, one more appropriate for a post-colonialised urban theory.

LOCATING MODERNITY

Commentators on Western history suggest the idea of modernity emerged there as a way of thinking about a secular, Renaissance time in contrast to the Middle Ages – a time that was then understood to be definitively past – and with an orientation to the future (Osborne 1992: 31, Fabian 1983). Modernity then arose as a form of historical consciousness,[2] a way of thinking about the present as new, qualitatively different from, if not better than, the previous ages (Osborne 1992). These ideas of the new are of course abstracted from the multiplicity of historical processes and times that coexist and shape any given moment. In this, the concept of modernity may well be a fundamentally flawed basis for historical analysis, gathering together under the banner of temporality a range of processes best understood in other registers. But the geopolitics of the term, as it came to translate chronology into spatiality, established its elemental role in figuring some places and people as out of time. As Osborne puts it, modernity, tied to ambitions of progress and via the discourses of colonialism, produced 'the idea of the *non-contemporaneousness* of *geographically diverse* but *chronologically simultaneous* times' (1992: 32 emphasis in original).

This enabled the identification of some people and some places as distinctively not-modern. Less an abstract identification of the simply new, this term of historical consciousness spatialised and racialised chronological categories. Although they were left out of this understanding of the modern and, instead, characterised as childlike, backward, primitive or traditional, the historical conjuncture of the modern did nonetheless depend upon these constitutive outsides (Laclau 1990). The people and places excluded from modernity were as often very close to home as they were dependent upon an exoticised and distant difference. The exclusion of women from the understanding of what it meant to be modern; or the association of an African diaspora or peasant and religious communities with the antithesis of modernity, has offered a basis for the emergence of an important critique of the concept of modernity from within the Western context. As Rita Felski writes in conclusion to her critique of gender and modernity,

> (T)he history of the modern needs to be rethought in terms of the various subaltern identities that have contributed to its formation [. . .] this involves a major fracturing and reshaping of established temporal schemata and periodizing structures. Received wisdoms about the aesthetics and politics of the modern will thereby be subjected to processes of contestation and revision, as the heterogeneous, often non-synchronous, yet intersecting modernities of different social groups come into view. The history of the modern is thus not yet over; in a very real sense, it has yet to be written.
>
> (Felski 1995: 212)

In response to these critiques, the history of the modern is in fact being rewritten – through feminist and minority politics within Western countries and also through a strong post-colonial critique of the consequences of modernity for places beyond the West. Hybridised and alternative modernities are being recounted and recorded in many different contexts, for the compromised form of modernity's conceptualisation does not mean that what is taken to be modern hasn't travelled well or hasn't encouraged enthusiastic affiliations from people perhaps far from the Western circumstances of its own ascribed origin (Askew and Logan 1994, Appadurai 1996). Indeed, diverse and quite different forms of modernity have emerged all over the world. Writing of Shanghai, Leo Ou-Fan Lee (2001) observes that during the early years of the twentieth century, 'despite the existence of the (foreign) concessions both the central government and Shanghai residents considered the city a Chinese metropolis' (2001: 115), and that with an 'unquestioned confidence in their Chinese-ness' (2001: 117), the residents of Shanghai 'turned Western culture itself into an other in the process of constructing their own modern imaginary' (2001: 116). Chinese residents of Shanghai appropriated and reimagined the many foreign components of the city as part of their own sense of being modern and being Chinese. Emergent Chinese nationalisms framed a cosmopolitan enthusiasm for ideas from other places, even colonial powers, without admitting the relations of cultural domination they might intend.

Even as the concept of modernity travels, though, it is often the case that the experience of modernity elsewhere is belittled, seen as simply an inauthentic copy, or a curious out-of-place phenomenon. Most importantly, the recounting of modernity's travels consistently replays a theory of its Western origins. Lee's narrative of Shanghai provides a stronger account of an indigenous modernity than most, and yet his narrative still locates much of the dynamism of that modernity in the engagement with the West.

Mitchell notes that in broadening accounts of the modern to include the production and circulation of modernity in and through other contexts, there remains 'a danger that instead of decentring the categories and certainties of modernity, one might produce a more expansive, inclusive, and inevitably homogenous account of the genealogy of modernity' (Mitchell 2000: 6–7). In his view, 'the discipline of historical time reorganizes discordant geographies into a universal modernity' (Mitchell 2000: 8). If we are to tell a story of different modernities, whose energy and dynamism have both diverse sources and uncertain outcomes, we need to start somewhere else, with other histories as well as other places. It is too easy to snap theoretical narratives back into the 'historical time' of Western modernity, rather than enable discrepant histories and modernities to challenge both the narrative and conceptualisation of these experiences.

Despite its many travels in academic analysis, then, including through the dislocations of post-colonial analysis, the concept of modernity itself has not been fundamentally dislodged from its Western origins, nor even truly pro-liferated or pluralised in the same way as the experience of modernity clearly has. And in understanding modernity, urban theory, like much cultural analysis, has been less mobile or agile and certainly less encompassing of diversity than the cultural practices of the cities it seeks to describe. This will become evident through the rest of this chapter as we explore how accounts of urban modernity have developed from the experiences of a few iconic Western cities and have circulated in such a way as to limit the acknowledge-ment of the modernity of different kinds of cities. It is one thing, though, to establish that accounts of modernity carry with them assumptions of an originary Western urban experience. But we also need to actively dislocate this privileged relationship between modernity and the West if we are to post-colonialise urban studies.

How, then, are we to ensure the dislocation of modernity from the West's self-proclaimed production of it? Dilip Goankar proposes that:

> One can provincialize Western modernity only by thinking through and against the self-understandings, which are frequently cast in universalist idioms. [. . .] To think with a difference that would destabilize the universalist idioms, historicize the contexts, and pluralise the experiences of modernity.
>
> (2001: x)

Mitchell agrees, and hopes that these tactics might deprive modernity of 'any essential principle, unique dynamic, or singular history' (2000: 12). We need to find ways to write modernity outside of the historical time of the West. The only way to do this is to ensure that there are grounds for appreciating and experiencing the modern without necessary reference to the West, or Western capitalism. This means disconnecting the social transformations and cultural valorisations indicated by theories of modernity from assumptions about progress, and from any fixed geographical referents. Decentring the West in theories of modernity means seeking to understand the sources and sites of social transformation wherever they may be and allowing for newness and innovation, along with their cultural valorisation, to emerge and exist anywhere.

This does not mean the search for a Holy Grail of autonomous cultural innovation or excitement at the new that is uncontaminated by Western or globalising processes. In the current era at least this may well be an historical impossibility. But it does mean that in the variable and discrepant encounters with novelty, wherever it comes from, the energy of innovation and the cultural practices of appreciation need not always be understood or marked as modern in a distinctively Western vein. Refusing the West's absorption of dynamism to itself, and insisting that innovation and creativity rests with all societies, it will be possible to relate accounts of diverse cultures of modernity. This could include diverse conceptualisations of the meaning of living in the present, and diverse accounts of important changes and developments in the lived experiences of people in many different places.

So, theories of modernity cannot be assumed to exist only in the West (Tsing 1993: 31–2) although, even outside of Western contexts, histories of colonisation and capitalist expansion may lead to 'Western' products such as religion, forms of state, consumer goods and cultural practices becoming 'inseparable from local concepts of modernity' (Li Puma 2001: 18). The weight of history here is profound, as '(Colonial) culture reinscribed world history in its own Eurocentric terms' and '(c)olonisers everywhere purported to export modernity, designating all others as "premodern" ' (Comaroff and Comaroff 1997: 27). But even so, it is crucial to acknowledge, along with the careful observations of scholars like Li Puma and the Comaroffs, that concepts and experiences of modernity around the world remain diverse, contested, contradictory and shaped by durable cultural practices, regional and local dynamics as much as by Western, colonial or global processes. To be attentive to these alternative and unpredictable trajectories of modernity would enable stories to be told of many different kinds of novel experiences and ways of engaging with the new – stories that would insist that newness emerges in all sorts of different places, with diverse consequences, and without necessary reference to the West. The aim, according to Goh (2002: 28), is to 'effect a theoretical disengagement from Western narratives of the modern'. Or, as Mitchell and Abu-Lughod suggest, 'we need to re-examine the way the histories of local subjectivities always tend to be written from the perspective of a global narrative of transformation' (1993: 80).

In many uses of the term, Timothy Mitchell writes, 'modernity is just a synonym for the West' (2000: 1). Assisted by the expansion and dominance of Western economic, political and cultural forms, the assumption that being 'modern' involves being 'Western' proliferates both in the academic literature, and in popular discourse as far apart as Papua New Guinea (Li Puma 2001) and Zambia (Ferguson 1999). It has been too easy to take the historically specific experience of Western colonialism and neo-imperialism as the template for appreciating the modern. If being modern is to be contemporary, to embrace change and dynamism, then the condition of modernity is present in every dynamic, changing society. This might be too general an appreciation of the term for some. But it quickly reserves any suggestion that there are static and unchanging societies. There are no societies without history and none outside of historical time. As Western understandings of modernity and tradition emerged as concepts alongside each other, the invention of tradition can only be a 'modern' phenomenon (Fabian 1983, Comaroff and Comaroff 1997).

A reconceptualisation of urban modernity might begin by interrogating Western forms of modernity. For these are as much a product of dynamics from beyond the West as its own ingenuity, including primitivist-inspired borrowings, actual resource extraction and political innovations learned in colonial governance (see, for example, Stoler 1995; Comaroff and Comaroff 1997). But western modernity's projection of itself as the generative source of creativity relies on forgetting these circulations and borrowings. The projects of modernity and modernisation in Western contexts have been transformed and also produced in their interactions with people and practices declared 'other' (traditional, primitive, backward) in the very process of defining the modern. Encounters with many different creative practices, individuals and groupings have contributed to the formation and reformulation of what it means to be modern in different contexts (Mitchell and Abu-Lughod 1993). The very promiscuity of Western modernity itself thus proposes a different, cosmopolitan cartography of modernity, one in which origins are dispersed, outcomes differentiated and multiple and the spatial logics those of circulation and interaction. Fixing an account of modernity in one place and time freezes the moment of innovation and belies the discrepant mobilities that have always underwritten innovation and newness.

In this cartography, cities do have a special place. Not because innovation happens there or because they are privileged sites of modernity – although there are often good reasons to consider cities to be innovative and closely linked to cultures of modernity. Rather, it is because in their diverse links to many different places and in their function as assemblages of social and economic relations, cities provide a model for the ways in which different, sometimes new, phenomena and experiences circulate to different places and accumulate a distinctive cultural meaning there. Western modernity, then, is only one moment in the astonishingly diverse circulations and productions of new things and new ways of being that are assembled in distinctive ways to produce different kinds of places and ways of understanding them. Cities everywhere perform this function of facilitating circulation, assemblage and interaction – of enabling diverse forms of 'modernity' to be imagined and practised.

This elementary form of 'spatial thinking' insists that those contingently Western cultural and material practices that have travelled the globe in the form of powerful, often dominant, social and economic relations are (a) not historically purely Western in origin, (b) only sometimes dominant and (c) definitely not the only style or theory of modernity around. The history and theory of modernity may circulate through and be shaped by the West, but it is not appropriate to reduce it to being simply 'Western'. Displacing the referent, West, from the experiences and theorisation of modernity is an historical possibility. Furthermore, as we will explore through this book, there are other models of innovation and modernity than the cartography that places the West at the origin and centre of the modern or the urban. Newness emerges and comes to be appreciated through many different routes and in diverse contexts.

The concept of modernity, then, bears the marks of its origins: the central core of my argument is that from the position of the post-colonial critic (Chakrabarty 2000), modernity can best be understood as the West's self-definition of itself (although not without its internal critics) in opposition to both the past and other, supposedly backward, societies. More than this, the idea of modernity has often been closely aligned with urban life although clearly the compass of the modern was also appreciated to reach far beyond any urban environment, in forms and routes of travel, modern agricultural practices, dams and resource extraction, for example. But in seeking to make sense of city life, for many decades urbanists have most often turned to align the city with the phenomenon of modernity. In the apparent capacity of cities to concentrate innovations and social vitality they, like no other places, have captured the dynamism of the contemporary, according to writers as diverse as Louis Wirth (1964) and Jane Jacobs (1965), Saskia Sassen (1991) and Ash Amin and Nigel Thrift (2002). A certain version of Western modernity, then, has had an intimate association with certain cities. Most importantly for our purposes here, this has meant that for certain kinds of cities to be thought of as modern, they have been routinely counterposed to 'tradition', antiquity, primitive or folk societies, or to societies assumed to be more 'backward' than those that have claimed the epithet of modernity for themselves.

In contrast to a view of the origins of modernity as Western, I want to insist that the sources of change and modernity in all societies are multiple and long standing. Reserving dynamism and invention for themselves, westerners designated other people as lacking in these qualities. But they were and remain wrong about this. While apparently European-centred processes of industrialisation and commodification might have sped up and globalised the cycles of certain kinds of inventiveness and change for a period, transformations of many kinds have always characterised human societies everywhere. Even a choice to defend tradition could be understood as a contemporary adaptation to the present – a modernity. Moreover, the dyna-mism of Europe itself depended upon and drew on many different parts of the world: much of Europe's self-acclaimed modernity was driven by events and actions elsewhere. For these, and many other reasons, it is no longer defensible to dispossess people and places of their creativity in the name of appropriating modernity to only certain sections of the world.

The next section turns to examine the ways in which Western theories of urban modernity have been produced through counterposing a distant world of tradition and primitivity to the rational practices of a specifically urban (and Western) modernity. A central fantasy of Western modernity – seeing Western cities as centres of rationality, dynamism and innovation – emerged, then, in a close relationship with assumptions concerning other kinds of people and places who were relegated to zones of backwardness and non-modernity.

URBAN MODERNITY AND THE PRIMITIVE

> The movement and migration of peoples, the expansion of trade and commerce, and particularly the growth in modern times of these vast melting pots of races and cultures, the metropolitan cities, has loosened local bonds, destroyed the cultures of tribe and folk, and substituted for the local loyalties the freedom of the cities; for the sacred order of tribal custom, the rational organisation which we call civilization.
>
> (Park 1967: 203).

The city has performed an important function in theorising modernity: it has coalesced and helped to make visible a certain range of self-descriptions for the West. Similarly, though, the idea of modernity has underpinned accounts of the city through the twentieth century. But in this double move, the idea of an urban modern has been intertwined with a counterpoint, the idea of the primitive. Running through so many accounts of the emergence and character of urban life in Western contexts has been the convenient category of the primitive, which has worked to create the fiction that these cities are modern. Here I turn to consider how cities have been conceptualised in some of the core texts of urban studies, especially the very influential Chicago School scholars Robert Park and Louis Wirth as well as the major source for their theoretical inspiration, the German sociologist Georg Simmel. As we will see, the enormously stimulating resources for thinking about cities that they contributed carried with them assumptions about the relationship between modernity and tradition[3] that continue to haunt Western urban theory.

For Robert E. Park, the city was 'the natural habitat of civilized man' (1952: 14). The idea of cosmopolitan urban civilisation reflected a state of freedom for the individual. This was a freedom that emerged, as he suggested in the above quotation, as 'those vast melting-pots of races and cultures, the metropolitan cities [. . .] substituted [. . .] for the sacred order of tribal custom, the rational organization which we call civilization' (1967: 203). In the city, Park thought, people were emancipated, their energies freed to pursue their individual place in the division of labour; rather than to assume their unthinking place in the undifferentiated collective that was primitive or peasant life (1952: 24) or being subjected to the 'control of nature and circumstance which so thoroughly dominates primitive man' (1967: 203).

The widened possibilities for human life that the city offered would, Park considered, lead to the disappearance of local and tribal cultures in the process of fusion in the city. To the extent that 'great cities have always been cosmopolitan', older kinship and cultural groups would be replaced by competition between people regardless of race or origin (Park 1952: 141). While this promoted economic efficiency and stimulated the ambition and energy of urbanites for 'new and strange inventions and achievements' (Park 1952: 141), the loss of customs and traditions also left new arrivals in the city bereft of the 'collective wisdom of the peasant community' – finding new ways of ordering social life was a key challenge of city life (Park 1967: 4). This upheaval and complexity of city life contrasted with an imagined primitive life that was seen as 'relatively stable' and in the bounds of which all individual and collective needs were met. Life centred around families and shared religious beliefs was being transformed in cities (1967: 15–16), and there was scope for individual talent to flourish once released from the suppressions of the intimate family circle or small community (1967: 18). The territorial restrictions of bounded primitive communities would be replaced by the worldwide influence and trading relations of great cities (1967: 227). In a still compelling phrase, he notes that:

> every great commercial city today is a world-city. Cities like London, New York, San Franscisco, Yokohama, Osaka, Shanghai, Singapore, and Bombay are not merely centres of a wide regional commerce. They are, by their position on the great ocean highway which now circles the earth, integral parts of a system of international commerce. They are way-stations and shopping centres, so to speak, on the main street of the world.
>
> (Park 1952: 133–4)

For Park one of the main consequences of the city in sociological terms, then, was the replacement of primary with secondary relations – family and kinship networks were replaced by the abstract relations of trade and work, as well as the even more distant interactions of daily city life. We will revisit this in Chapter 2 at some length, as Park's hypothesis about the indifference, or blasé attitude of city dwellers was the subject of many comparative studies of urbanism that we will review there. For now, it is important to note that the idea of the folk, or the primitive, captured those qualities with which the city was no longer associated: in small communities the most intimate and real relationships of life were 'practically inclusive', intercourse conducted 'largely within the region of instinct and feeling' and social control a product of personal accommodation, 'rather than the formulation of a rational and abstract principle' (Park 1967: 33). To the 'city' then, rationality, thought, distanciated social relations; to the 'primitive', intimacy, feeling, sentiment, instinct and the absence of reason. The achievements of urban modernity were not overwhelmingly positive, though, and the dominance of secondary relations and loss of primary communities in the city was associated with a range of social and moral ills.

The primitive came to hold a lot of Park's attention as he reached back into literature from the nineteenth century to make sense of the world-historic changes he associated with city life (for example, in 1967: 230). Later, Hausner, a strong critic of the ethnocentricism of the folk-urban continuum, was to note, 'In some respects, these ideal-type constructs represent an admixture of nineteenth-century speculative efforts to achieve global generalization, and twentieth-century concern with the integration of knowledge for general education purposes, as a result of which integration is often achieved of that which is not yet known' (1965: 514). In so far as these contrasts between cities and rural/traditional/folk/primitive others were largely 'imaginative' achievements, they drew richly on the received wisdom of their times (Hannerz 1980: 64).

'Primitives' were a counterpoint to Park's discussions of collective and public life in cities, the function of the crowd, as well as to his analyses of community, sociability, migration and emotions (Park 1967: 196, 225, 230). The observations of colonial scholars sustained his understanding of how distant, primitive societies contrasted with emergent urban social relations. And through the eyepiece of the category primitive, Park assessed the transition to life in cities for immigrants from the 'back' areas of Europe, from the plantations of the American South, as well as the 'marginal' experiences of Jews and Gypsies (1967: 22, 24, 116, 119). The analytic of the primitive sustained most of Park's major conclusions about life in the 'modern' world, and for him (and many others who followed) 'primitive' and folk cultures were generally considered to have cultural practices diametrically at odds with those of 'modern man'. By the mid-1960s, when Park's account of urban life had been well established and much researched, the idea of the folk or primitive society had accreted a long list of characteristics, described here by a critic of his analysis as:

> small; isolated; nonliterate; homogenous; strong sense of group soli-
> darity; simple technology; simple division of labour; economically
> independent; possessing 'culture', that is, an organisation of con-
> ventional understandings; behaviour strongly patterned on a con-
> ventional basis – traditional, spontaneous, uncritical; informal status;
> no systematic knowledge – no books; behaviour is personal; society is
> familial; society is sacred; mentality is essentially personal and emotional
> (not abstract or categoric); animism and anthropomorphism manifest;
> no market, no money, no concept of 'gain'.
>
> (Hausner 1965: 505)

Being alive to the cultural vitality of cities for Park and many others who both influenced and came after him involved the entrainment and often the denigration of a range of cultural differences as 'primitive' or 'traditional'. Of course this optic – counterposing the civilised world with primitive cultures – was one that was common at the time, and the point here is not to dismiss the intellectual labour of generations of scholars because of it. And certainly in his wide-ranging writings on social life in cities, Park was well aware that close and stable ties existed in the city as well as in traditional communities (Hannerz 1980: 25). But then again, it is also possible to argue that without the primitive, the sketch of city life which they offered could not have been sustained or perhaps even generated. It might not have been so easy to portray cities as special sites of creativity and dynamism if the fiction of the primitive as static and unchanging had not cast this narrative into careful relief, if the primitive had not been available to hold in abeyance what theorists imagined the city not to be.

Georg Simmel, inspiration for much of Park's insights on city life and the sociology of the 'modern', was a little more careful in the way he drew distinctions between metropolitans and their others. People in small towns, villages, families, religious and kinship groups – these were all specified as counterpoints to life in the city. But Simmel's analyses are somewhat more dialectical than Park's. Less concerned to establish an understanding of the world-historic role of cities in Western civilisation, Simmel tracks how the 'sensory foundations' of psychic or social life vary between large cities and other kinds of places. In his famous essay from 1903, 'The Metropolis and Mental Life', the primitive figures as a comparator to the modern fight for individuality in the face of the collective (as primitives struggled with nature for their 'bodily existence') (1971: 324); and as an anchoring point in 'one of the great developmental tendencies of social life as a whole'. He saw this as involving the progressive development from small enclosed groups, which are both isolated and permit only a slight area in which individuals can develop their own qualities and activities, to large, extensive groups with mutual relations and connections beyond the group and a substantial measure of individual freedom, associated with the emerging division of labour (1971: 332).

Like Park, then, for Simmel 'freedom' of the individual is a great achievement of the city and of its spatial extension, wider connections and enlargement of public life. But Simmel casts this development of individual freedom as having emerged from the communal sociabilities of a wide range of different contexts and groupings both within and beyond the city. 'Political and familial groups, political and religious communities [. . .] the State and Christianity, guilds and political parties and innumerable other groups', he suggested, have all 'developed in accord with this formula', that is, they began as small enclosed groups and are transformed into more open and rationalised social relations. The 'others' to modern city life in Simmel's account encompass the figure of the primitive, both elsewhere and in a developmental past, but also include contemporary European social forms: 'The most elementary stage of social organisation which is to be found historically, as well as in the present' (1971: 332). He also identified these communal social relations as present in earlier forms of urbanism: of cities in the Middle Ages he notes that the 'barriers were such that under them modern man could not have breathed' (1971: 179).

Perhaps most importantly to Simmel's dialectical imagination, a contrary tendency to this emergence of individual freedom is also evident in the history of the city. He observes the tendency for collective metropolitan achievements to overpower individuality – the individual might come to feel 'a mere cog in an enormous organisation of things and powers which tear from his hands all progress, spirituality and value' (1971: 184). This diminution of individuality heightened investments by some people in distinctive, fashionable forms of self-presentation, so that they might stand out in the crowd and be recognised. This stimulates strong investments by urban dwellers in a search for new and distinguishing fashions and novelties. In contrast, Simmel notes, 'Among primitive races fashions will be less numerous and more stable because the need of new impressions and forms of life, quite apart from their social effect, is far less pressing' (1971: 302).

To Simmel, then, the city provided an Hegelian-style arena in which the 'world history of the spirit' could work itself out in a battle between these contradictory urban tendencies, of enabling individual freedoms, yet providing obstacles to (and thus stimulation for) the preservation of individuality in the forces of rationalisation and industrialisation (Frisby 2001: 158). Still, lurking behind these theorisations of both individualism and individuality, and easily excavated by the American-based scholars of the Chicago School steeped in a world of racialised difference, was the figure of the primitive and the folk, peasant and 'backward' cultures of various 'closed and bounded' societies.

It was Louis Wirth's more static and categorising account of the sociology of cities that was most widely read through the twentieth century. His essay on 'Urbanism as a Way of Life' drew together the insights of both Park and

Simmel in a strong statement on the sociological distinctiveness of life in cities. Writing, as did Park, in a time of very rapid in-migration and growth in Chicago, the proximity of European peasant immigrants immersed in folk culture and struggling to adjust to life in this US city (and for Wirth this was a personal journey he had made as a young man) posed the question of the relation between 'folk' and city life in stark ways. We could hypothesise that this might have prompted some anxiety and thus provoked these scholars to make extremely clear and strong statements as to what being in the city entailed and to distance themselves from the 'backward' practices of recent migrant communities. Certainly, their assessment was that the cultural practices of 'folk' traditions such as those brought to Chicago by East Europeans, Italians, Germans or 'Negroes' and Jews would not persist in this new environment. Being in the city was NOT to be folk, traditional, primitive, and so on. Any evidence of these cultural practices was merely transient and was not to be associated with the cultures of city life.

Inspired by Thomas and Zaneicki's (1927) *The Polish Peasant in Europe and America*, Wirth assessed migrants to America's cities as experiencing social disorganisation, following the breakdown in norms and customs that had sustained order in Europe's rural areas. Again, it is the 'primitive' and folk culture that highlights the distinctiveness of the move to the cities: conflicts between norms, Wirth writes, 'are rare in stable, compact, and homogenous societies, as an abundance of literature from primitive and folk societies indicates' (1964: 47). The dynamism and alienation of city life and the characteristic indifference that it fostered were compared favourably to the stability and intimacy of life in other kinds of places and in other kinds of cultures. In contrast, the multiplication of norms in cities in a context of migration was understood to lead to conflicts and tensions and highlighted the need for new, rational and collective forms of social organisation (1964: 49). The ghetto ('be it Chinese, Negro, Sicilian, or Jewish' (Wirth 1964: 8)) forms an internal counterpoint to these emergent new forms of modern city life:

> for, with all its varied activities and its colourful atmosphere, the ghetto nevertheless is a small world. It throbs with a life which is provincial and sectarian. Its successes are measured on a small scale, and its range of expression is limited [. . .] it is a cultural community and constitutes as near an approach to communal life as the modern city has to offer.
>
> (Wirth: 1964: 98)

For Wirth, the ghetto, like other forms of traditional and communal life in the city, was destined to disappear in the urban melting pot.

Building on these observations Wirth seeks to develop a general sociological theory of urbanism, one that would not be determined by 'any specific locally or historically conditioned cultural influences' (1964: 64). It would only 'denote the essential characteristics which all cities [. . .] have in common' or, as he put it, 'at least in our culture' (1964: 62). With his understanding of urbanism as 'That complex of traits which makes up the characteristic mode of life in cities' (1964: 66), he works with his most general definition of cities as 'a relatively large, dense, and permanent settlement of socially heterogeneous individuals' (1964: 66). From these spare beginnings, Wirth weaves an account of what it is to be urban. The 'primitive' plays a slightly less obvious role in this essay, but many of the concepts that Wirth draws on depend on the earlier works of Simmel and Park and imply a primitive counterpart. He repeats their concerns with the absence of kinship bonds, neighbourliness and personal mutual acquaintances in cities (1964: 67) and notes the personal and emotional qualities of intimate groups, the spontaneous self-expression, morale and participation that follow from integrated societies (1964: 68), as well as the simple hierarchical arrangements and concentric social relations of 'rural communities of primitive societies' (1964: 72), alongside their rigid caste distinctions. By contrast, in the city he observes the presence of freedom (from all these things), as well as the predominance of rationalisation and secularisation (1964: 71), greater differentiation of social structure and cosmopolitan sophistication, sharpened income and status distinctions (1964: 77) and more fragile, complicated forms of human association.

Wirth's hope for a unified body of general theoretical knowledge about urban sociology rested, once again, on a social world assumed to be cleansed of a range of social and psychological processes that were consigned to the categories of the primitive and the traditional. He did observe, though, that a measure of historical continuity in the production of cultures and social relations meant that 'our social life bears the imprint of an earlier folk society', making the discontinuity between traditional and urban ways of life a little less stark than at times he might suggest (1964: 59). And he also noted the strong influence of urban ways of life on cultures outside the city, in rural areas, for example (1964: 62). These nuances aside, the city and the country, modern city dwellers and the primitive, were cast as opposites: an opposition that sustained a restricted account of the ways of city life.[4]

Simmel's influence on Wirth is reflected not only in the borrowed concepts of a blasé outlook or rationalised urban way of life; it is also clearly visible in Wirth's easy borrowing of both Simmel and Park's construction of the urban through the figures of the folk and the primitive. Thus, cities could be identified as places where people were rational, modern, dynamic, individualised and involved in stratified industrial forms of economic relations, at least partly because the figure of the primitive placed characteristics and practices 'other' to these ways of being urban elsewhere, outside cities – specifically 'back then' (in traditional times) and 'over there' (in backward places).

It was one of Simmel's other admirers (and critics), Walter Benjamin, writing at much the same time as Park and Wirth, who engaged more critically with these assumptions about the relations between urban modernity and its 'others', at least partly because like Simmel he was a profoundly dialectical thinker. Benjamin explores the concept of Western modernity by postulating a dialectical relationship between conceptualisations of the modern (here and now) and the idea of tradition, antiquity, or the primitive. Benjamin offers a critical account of this relationship between modernity and tradition as produced within modernity itself. In his writings on cities, he also begins to sense how this dialectic between modernity and tradition might work through different kinds of cities and, in doing so, introduces another effect of the concept of the 'primitive' in urban studies: its ability to hierarchically order different cities in relation to their varied achievement of modernity.

A DIALECTICS OF MODERNITY AND TRADITION

Whereas the Chicago School writers conveyed a strong sense that cities represented an advance on life in traditional or primitive societies, Walter Benjamin's philosophy of history explicitly refuses a theory of progress. He stated that he wanted his work 'to demonstrate a historical materialism which has annihilated within itself the idea of progress' (1999a: 460). It is not surprising, then, that Benjamin's work does not establish a general account of urban life against the concept of the 'primitive'. Instead, he proposed a dialectical materialism interested in the specific 'constellations' that the era of the historian ('the time of the now') might forge with 'a definite earlier' time (1999b: 255). Rather than being part of a historical continuum, Benjamin was concerned to 'blast' historical objects out of any supposed causal sequence of events and to reframe them within a 'now of recognition' – within a dialectical image, or the dialectic at a standstill (Buck-Morss 1989: 219). The relationship between 'antiquity' or 'tradition' and the modern was not one of progression, one following the other in 'homogenous, empty time' (Benjamin 1999b: 252). Rather, for Benjamin, notions of antiquity and tradition are born in the time of modernity (1999a: 465) and are often attached to contemporary projects through political gestures that seek to lay claim to the stature and mythology of earlier times. Or, as we have seen in the case of the Chicago School theorists, the past (and elsewhere) is mobilised in the present to figure and hold at bay qualities that, they argue, are not characteristic of the here and now. A dialectical imagination, then, insists on the co-presence and mutual interdependence of concepts of modernity and tradition (see also Kusno 2000, Crysler 2003, King 2004: 72).

Most importantly, Benjamin considered the interplay between modern innovations (new technologies, new commodities and inventions) and tradition – whether staged through scholarship or found in cultural artefacts or the urban environment – to be dynamic and potentially transformative. He saw the phantasmagoria (the fantastic, imaginary cultural representations) of modernity as a site that potentially exposed the range of alternative possibilities for organising social life. In seeking to construct this revolutionary critique of contemporary cultures, he was drawn to thinking about the form of commodities and also about the supposedly rational, ordered and technologically sophisticated nature of city life. Both of these objectively modern phenomena carried with them memories, dreams and fantasies from the past. For commodities this often took the form of advertising, which frequently spoke to consumers' wishes to recapture a utopian past through buying or consuming the object; in cities, the ruins of past urban dreams littered the landscape of cities bringing into the present the unfulfilled hopes of past times. Benjamin saw both of these dynamics as having the potential to draw city dwellers to a realisation of the need for revolution; the bankruptcy of current promises and power relations could be exposed through analysing these dialectical images. The ambiguous use of tropes of tradition, or of a utopian past to ascribe novelty and excitement to newly invented forms of technology, commodities or buildings was, for Benjamin, an especially important starting point for exploring the contradictory dynamics of urban and modern life:

> But precisely the modern, *la modernité*, is always citing primal history. Here, this always occurs through the ambiguity peculiar to the social relations and products of the time. Ambiguity is the manifest imaging of dialectic, the law of dialectics at a standstill. This standstill is utopia and the dialectical image, therefore, dream image. Such an image is afforded by the commodity per se: as fetish. Such an image is presented by the arcades, which are house no less than street. Such an image is the prostitute – seller and sold in one.
>
> (Benjamin 1999a: 10)

The same dynamics of the dialectic at a standstill he observed in the ruins of past eras, still present in the city. His exploration of the nineteenth-century Parisian arcades was, at one level, an attempt to evoke through these still present, if now rather ghostly and dreamlike, elements of the city the alternative futures they had once spoken of to contemporary society. A future where the public and the private might be intermingled and where the utopian fantasies through which technological innovations were often represented harkened back to a pre-class and pre-commodity society of plenty (Buck-Morss 1989).

These representations brought opposing possibilities, alternative futures crashing together and in illuminating these Benjamin hoped that he would 'wake' the sleepwalkers of the modern city up to a revolutionary consciousness about the problems of the present city (Pile 2005). In the moment of awakening, and through the recognition of a 'constellation' of present and past, with the traditional imbricated in cultures of the avowedly modern, Benjamin saw not just the backward or irrelevant left-behind elements of social life, but the possibility for a transformative politics opposed to commodity fetishism and capitalist forms of urban development. According to one commentator, he was exploring 'The speculative power of ruins and the

outmoded [. . .] arrested in the threshold of any number of possible futures'
(Caygill 1998: 135). The traces of the lost futures of the past, buried in the
remnants of the city, offered a way into imagining different possible futures
for the present.[5]

Benjamin's dialectical imagination can offer us some resources, then, for
challenging the dependence of theories of urban modernity on simple contrasts with the primitive and tradition. In his wake, we might search for
dialectical images that stage the encounter between tradition and modernity,
and that also challenge easy assumptions about the alignment of modernity
with the city and tradition with 'back then and over there'. We will turn to
this in the following section. Before we do that, though, it will be useful to
explore how some of Benjamin's writings on the cities he visited alert us
to the consequences of theories of modernity for how we understand the
relationships amongst different cities. If the Chicago School's theories of
urban modernity implied a geographical imagination in which cities are contrasted with rural areas, they also had consequences for how modernity and
tradition were thought to be distributed across different cities. And in relation
to this issue, Benjamin's writing is a little ambivalent, reflecting the depth of
the colonial unconscious, perhaps, in his intellectual environment.

Early in his period of thinking and writing about urban culture in Berlin
and, more famously, in Paris, Walter Benjamin visited both Naples and Moscow and wrote two short essays (the first with Asja Lacis). Each of the cities
was different, in different ways, from the northern European cities with which
he was familiar. The alternative forms of urbanism he encountered in Moscow and Naples offered a way of bringing into focus the historical specificity
of the European urbanism with which he was familiar.

In the essay on Moscow he comments, for example, that 'all Europeans
ought to see, on a map of Russia, their little land as a frayed, nervous territory far out to the West' (1979: 196). He notes at the beginning of this essay
that '(m)ore quickly than Moscow itself, one gets to know Berlin through
Moscow', and that knowing what was going on in Russia, 'one learns to
observe and judge Europe' (1979: 177). Caygill suggests that these reflections
on different cities were not only 'geographical but also chronological' (1998:
118). Moscow in 1927 reflects a possible future, a city in the throes of revolutionary transformation. And Naples the 'traditional' past or, as Susan
Buck-Morss expresses it, 'the myth-enshrouded childhood of Western
civilization' (1989: 25). Both agree that Benjamin's engagement with the Paris
Arcades Project was informed by his analysis of these cities: 'It was only
through the analysis of these other urban experiences that Benjamin was able
to understand his own' (Caygill 1998: 128). The similarities that he identified
between the Parisian arcade (where the boundaries of public and private
spaces were transgressed) and the porosity of Naple's street life,[6] where the
realms of public and private intermingled, is one key indicator of this (Caygill
1998: 132). Buck-Morss (1989: 27) also suggests that the Naples experience
had 'key methodological import' for Benjamin's later work in establishing

the images gained in experiencing the city as 'objective' expressions of the materiality of city life rather than purely subjective impressions.

But, in his later work on Paris, Benjamin never developed any sustained comparative observations about different cities, which has invited the easy extension of his specific assessments of urbanism in Paris and to a lesser extent Berlin, Naples and Moscow to a more universal relevance (see, for example, Savage 1995, Amin and Thrift 2002). In the Arcades project, though, he made a note which suggestively indicates an elaboration of his concern to avoid such universalising judgements, criticising notions of historical 'progress' or decline in relation to different kinds of cities. It takes the form of a pithy reflection, a kind of 'note to self':

> The pathos of this work: there are no periods of decline. Attempt to see the nineteenth century as positively as I tried to see the seventeenth, in the work on *Trauerspiel*. No belief in periods of decline. By the same token, every city is beautiful to me (from outside its borders), just as all talk of particular languages' having greater or lesser value is to me unacceptable.
> (Benjamin 1999a: 458)

Here he is intimating a desire to avoid assessing the value of different cities against some external criteria and, as in his study of languages, he is unwilling to ascribe a hierarchical valuation of one city over another. And yet, in the range of metaphors that Benjamin drew upon to address the differences amongst the cities he wrote about, he includes images of 'backwardness', primitiveness and the interpenetration of country and city to distinguish between the northern cities he usually lived in and wrote about, and these other cities that he had visited. Here the primitive is spatialised, but in a new kind of way, associated with certain cities, rather than simply with past times, rural or exotic places. He draws these other places and times into his analysis of these cities themselves.[7]

Creeping into his accounts of these cities, then, are some other, quite distant places: 'In his use of time', he observes, 'Russia will remain "Asiatic" longest of all' (1979: 190); advertisements are appealing, in his opinion, to 'primitive' tastes; and Moscow in many places resembles a Russian village, 'rather than the city itself' (1979: 202–3). In Naples, he finds a useful comparator in the African kraal:

> What distinguishes Naples from other large cities is something it has in common with the African kraal; each private attitude or act is permeated by streams of communal life. To exist, for the Northern European the most private of affairs, is here, as in the kraal, a collective matter.
> (1979: 174)

Of course, the ascription of tradition as modernity's other is not always an overwhelmingly negative feature of social life in supposedly traditional societies. For around the time Benjamin was writing on cities a form of primitivism was very popular (Clifford 1989), especially amongst avant-garde artists and the surrealists, whose work he was familiar with. The primitive, often figured as African, in the realm of art, design, performance and litera-ture, was considered potentially transformative of a decaying European culture; a culture imagined to be desperately in need of invigoration (Szondi 1988: 24). Along with other sources of dynamism and possible revolutionary transformation, including the realm of the unconscious (especially for surrealists) and the outmoded (Benjamin 1979: 229), the primitive was a sign of purity, of uncultured and thus more revolutionary and truly human experience (de Certeau in Donald 1999: 72, Morton 2000). Of course this is to assume that those who are designated by the 'moderns' as 'primitive' are lacking in culture and sophistication, but it is important to recall that this was also imagined by these writers as a positive feature.

In these primitivist manoeuvres, then, Benjamin was seeking ways of critiquing and, if possible, transforming the exploitative and class-divided capitalist urban society in which he was living. One of the key themes of the Moscow and Naples essays was the porosity of city life and the interpenetra-tion of different elements of the city. By contrast with the northern cities he knew well, in Naples, Benjamin was impressed with the porosity of urban space: 'Porosity is the inexhaustible law of the life of this city, reappearing everywhere' (1979: 171). He and Asja Lacis commented extensively on the intermingling between public and private life, which they felt explained the dynamism of social life there. Commenting on some of the local building materials he noted: 'As porous as this stone is the architecture. Building and action interpenetrate in the courtyards, arcades and stairways. In everything they preserve the scope to become a theatre of new, unforeseen constellations' (1979: 169).

The term, 'constellations', is freighted in his philosophy with the potential for revolutionary awakening (Caygill 1998), and its use here indicates that he was trying to convey a sense of the hopefulness embedded in this description of the different urbanism he encountered in Naples. From the mobility of life in this city, he is suggesting, the possibility of new alignments and inter-sections emerging was palpable.[8] The porosity and transitivity of Moscow's urban life was similarly observed, for example, in the busy street life and the shifting experiences of space associated with the extreme seasons. But, for Benjamin, the 'astonishing experimentation' of the revolution there had made of the city a site of constant mobilisation – physical, social and political reorganisation was insistent (1979: 186). In this sense, Moscow offered a view of a post-bourgeois urban society: 'the privacy characteristic of bourgeois cities had been "abolished" by Bolshevism' (1979: 187).

The experience of transformation in Moscow, like the ongoing vitality of Naples, was figured by Benjamin through the continued coexistence of traditional and non-urban elements in the city, in stark contrast to the more thorough-going modernity of northern cities. In Moscow, he suggested, visitors had to adapt themselves to many things, including 'the complete interpenetration of technological and primitive modes of life' (1979: 190), the distinctive forms of timekeeping, and the 'lifeless wasteland' of the roof-tops 'having neither the dazzling electric signs of Berlin, nor the forest of chimneys of Paris, nor the sunny solitude of the rooftops of great cities in the South' (1979: 200). Moscow, Berlin and Paris were drawn into an imaginative, theoretical relationship in which Moscow and Naples were placed more on the side of the primitive and the archaic; Paris and Berlin on the side of modernity.

Alert to these differences, for Benjamin Moscow provoked a sense of dis-orientation as a result of the differences between city life there and in other cities that he had lived in and visited. Szondi suggests that Benjamin's encounters with 'things foreign' did not 'lure' him into self-forgetfulness – as perhaps might follow from a traveller's pure celebration of exotic difference; rather they encouraged him to see himself 'with an estranged vision' (Szondi 1988: 21). For Benjamin, Szondi suggests, wandering around strange cities was 'in space what memory, which seeks hint of the future in the past, is in time' (1988: 22). These different cities, then, jumbled up experiences of past and present, self and other, offering opportunities for reimagining oneself, one's relation to the past and to the cities one lived in. Benjamin certainly seems to have encountered himself and his own biography and context anew through his visits to these cities. In Moscow he suggested he felt as if he had entered a 'childhood phase' (1979: 179), where he had to re-learn how to interpret the city, or that while riding along in the little sleighs they comingled with people, horses, stones, and he felt touched and belittled by them, which was quite different from his experience of gazing down a street in the clear perspective of Haussmanised Paris (1979: 190–1). But, beyond these personal reflections, and following Szondi (1988), I suggest that Benjamin also drew on a spatialised imagination of these other cities as exotic and different to make sense of the images of city life he gathered there.

The quietness of Moscow drew his interest because of the winter snow, the silence of street traders and because, in his view, the traffic was more 'backward' than that of northern European cities. In this silence, Moscow is likened to a village, 'deep in the Russian interior' (1979: 203). This image of a city interspersed with the countryside, although surely one of inter-penetration and porosity, is here associated with a lack of cityness, rather than its future transformation: 'Nowhere does Moscow look like the city itself' (1979: 203). In Naples, too, the vitality of city life is traversed by a similar kind of judgement:

> This is how architecture, the most binding part of the communal rhythm, comes into being here: civilized, private, and ordered only in the great hotel and warehouse buildings on the quays; anarchical, embroiled, village-like in the centre, into which large networks of streets were hacked only forty years ago.
>
> (Benjamin 1979: 169–70)

These two cities are also brought together to form the basis for comparisons with each other (in Moscow 'Begging is not as aggressive as in the South' (1979: 184)), and to suggest how both are different from Berlin or Paris – in their timekeeping,[9] or their busy street life,[10] for example. Benjamin comments on this from a position of the visitor who has returned to Berlin:

> Returning home, he will discover above all that Berlin is a deserted city. People and groups moving in its streets have solitude about them. Berlin's luxury seems unspeakable. And it begins on the asphalt, for the breadth of the pavements is princely. [. . .] Princely solitude, princely desolation hang over the streets of Berlin [. . .] But what fullness has this street that overflows not only with people, and how deserted and empty is Berlin. In Moscow, goods burst everywhere from the houses, they hang on fences, lean against railings, lie on pavements.
>
> (1979: 178)

One senses the potential for Benjamin to draw some conclusions about the existence of a diversity of urban experiences, and to become aware of the very particular form of urbanism commonly described as characteristic of Paris and Berlin, in the writings of Simmel and others. If the 'crowd' and the 'shocks' of life in the city seem so important to writers on these cities (and are so central to Benjamin's work), and yet by comparison with Naples and Moscow, the streets of these cities are deserted, a different range of questions might be provoked about the experiences of city life. If streets are so empty, one would be forced to wonder why people in these places found the city quite so shocking and overwhelming, as was often recounted. And one would be drawn to ponder how it was that Benjamin sensed the vitality and enjoyment of city life in Naples with its many diverse happenings and crowded situations, rather than emphasising the shocking, individualising effects of city crowds. We'll reflect more in the following chapter on these aspects of Western accounts of urbanism, especially the question of whether the city is a site of alienation in response to excessive emotion, or a site of sociability and emotional interaction. We will be able to extend Benjamin's very preliminary comments on the diversity of urban experiences in different kinds of cities through an engagement with the work of slightly later mid-twentieth-century urban anthropologists who pursued a comparative account of urbanism in cities around the world, in direct confrontation with some of the claims of Simmel, Park and Wirth.

In this regard, it is the way in which Benjamin interweaves the 'traditional' and the 'modern' in his effort to capture a sense of the differences between cities that is important. For in doing this, Benjamin places the primitive, not on the side of the non-urban, but on the side of *different* urbanisms. In his account of Parisian urbanism, the historical dialectic with tradition embedded in the phantasmagorias of ruined modernities holds the seeds of a revolutionary consciousness for Benjamin. Perhaps the spatial mixing of these categories – modernity and tradition – in his accounts of 'different' cities hints at the possibility of a similarly provocative dialectic at a standstill in urban theory? Such a dialectic, staged as a result of the co-presence of modernity and tradition within cities, intimates the possibility of transforming the intellectual phantasmagoria of urban modernity. This phantasmagoria as we have seen, has been parochial and restricted through its exclusion from the realms of city-life features commonly ascribed to tradition or to the 'primitive'. How cities are imagined and theorised would be significantly transformed if these excluded qualities were understood to be both within the city and contributing to the production of urban modernity.

I do not think this was Benjamin's conscious intention, although, as we have seen, it would not be inconsistent with his wider historical analysis. And Kraniauskas (2000) offers a very suggestive reading of Benjamin's lack of attention to the colonial relations shaping many of the Parisian urban processes he was writing about (King 2004: 69). This absence is, Kraniauskas proposes, redeemed in the unconscious portrayal of a colonial presence in some of Benjamin's recorded dream images in *One Way Street*. Just as Benjamin 'forgot' the colonial (despite personal interest in 'Aztec' history and language, for example), so too have the literatures that have built on his work (Kraniauskas 2000: 141). But Benjamin's dream life failed to completely repress the colonial present. He dreams of a Mexican shrine (a church steeple) being uncovered in a Weimar marketplace and wakes up laughing, as the name he gives to the period (*Anaquivitzli*) embodies the German word for joke (*witz*) (Benjamin 1979: 60). The scene may stage a strange surrealist juxtaposition: the buried church, the marketplace. But for Kraniauskas, Benjamin's conversion of the dream into a waking joke diverts his attention from the ways in which the Mexican church, underground, stages the dependence of European capitalism and development on colonialism and primitive accumulation (Kraniauskas 2000: 148). And although for Benjamin the waking moment – the 'threshold' between the dream and waking life – is often figured as a moment of revolutionary optimism, the dream image here is resolved into a joke, covering over the unconscious truth about the buried colonial experience. Remaining unconscious for Benjamin and never directly explored, the dependence of Western urban modernity and accounts of the modern city on spatially (rather than historically) distant others remains to be excavated. Like the essays on Naples and Moscow, the dream reveals the coexistence in the city of 'modernity' and its others – colonised, distant, traditional – even if for Benjamin this remains a buried truth.

However, these dynamics were soon to come to the fore in urban scholarship. As researchers showed that social and cultural practices figured in urban theory as 'modern' and 'traditional' were both present in different cities around the world, an important challenge was raised to the Chicago School's alignment of modernity with the city. Not very long after Benjamin was writing, then, the colonial unconscious of urban theory was carefully and deliberately exposed. The 'constellation' of diverse urban experiences captured in the outpouring of anthropological and sociological representations of 'tradition' and modernity in cities around the world mark a moment in urban theory when a more cosmopolitan account of city life was, for a while, perhaps possible.

CITIES TROUBLED BY TRADITION

Benjamin's revolutionary interest in the dialectics of modernity and tradition was not explored by American sociologists, although they were, through Park, also inspired by Simmel's dialectical urbanism. Instead, the Chicago School legacy and reading of Simmel counterposed, in a very undialectical fashion, a distinctive form of urban attitude to a 'folk' culture that was imagined to be both past and elsewhere. This folk culture was figured as elsewhere in an immediate sense – in the countryside surrounding the large European and American cities on which writers like Park, Burgess and Wirth based their analyses (Wirth 1964). As we have seen, though, it was also often associated with 'primitive', or 'savage' people. Located both in the distant past of the West and also in the contemporary 'elsewhere' of colonised places, the figure of the primitive played a constitutive role in what could be said about 'cityness' in Western contexts, notably, enabling their identification as dynamic, modern, innovative, rational. But, and this is the challenge that urban anthropologists brought to the Chicago School theorists, if the 'primitive' was in the city, understandings of cities as sites of a particular version of modernity would have to be reconfigured.

A number of Western anthropologists and sociologists were turning from the 1940s to explore cities in other, mostly poorer countries around the world. As Hilda Kuper explained:

> The anthropologist, whose field itself is changing from preindustrial, small-scale, folk communities into industrial, large-scale, urban societies, is being forced to re-examine his techniques of research and his theories of interpretation. At the same time sociologists, confronted with data from non-Western urban societies, as well as from recent detailed studies in America and England, are querying previously accepted generalisations based on limited studies of Western cities.
>
> (1965: 10)

For these writers, what Wirth and others had been calling non-urban 'traditional', 'folk' or 'primitive' culture was more immediately present *within* cities. It was in fact quite evident in some of the experiences of recent migrants to the cities they studied (Abu-Lughod 1961), in the distinctive 'traditional' form of 'primitive' (but contemporary) cities (Miner 1953), or in the surrounding areas that were often very closely connected to city life (Mitchell 1973).

As poor 'tribal' people in Africa, for example, moved to the city, 'where anthropologists followed them' (Devons and Gluckman 1964: 13), anthropologists observed that they brought their 'folk' cultures to the city. Sometimes very strong links to their home places were sustained, where supposedly 'rural' or 'tribal' culture predominated (Mayer 1971). If, for Wirth, it was clear that urban culture was not necessarily confined to the physical area of cities, but that as cities spread their influence across the countryside and beyond, features of urban culture could pop up almost anywhere (1964: 64), for anthropologists the nature of the city itself was powerfully shaped by a range of social processes, which the Chicago School theorists had declared un-urban. These comparative studies of the city troubled existing accounts – and were very self-conscious about doing so. (The following chapter explores this in greater detail.)

The difficulty for scholars of cities shaped by the 'elsewhere' of so-called traditional, or 'primitive' societies was that, from the point of view of Western urban theory, the subjects of their study seemed almost by definition to be not-modern, and to be not-urban. A revision of the conceptualisation of the urban was called for. A generation of anthropologists had lengthy, even angry, debates with the received wisdom of Western urban scholarship. Writing in 1973 to introduce a collection of urban anthropological writings from around the world, Aidan Southall noted that 'It is also important for the Western nations to study with care and humility the new urban forms emerging in the non-Western world, to see if they offer any lessons as yet unthought-of in Western ethnocentric assumptions' (1973: 4). One of his concerns was about the strong continuities, rather than stark oppositions, between cities and countryside (1973: 9). Also, in contrast to the overwhelming and alienating experiences of city life described by Chicago School sociologists, he observed that many people in cities were developing tightly knit solidarities, rather than blasé attitudes of indifference (1973: 13). Indigenous and very different forms of urban living, closely associated with agricultural and communal practices, were also observed and documented (Lloyd 1973).

Oscar Lewis, whose famous 'culture of poverty' thesis had mixed receptions, made a strident critique of the associations that had been drawn between folk and 'primitive' culture, the tendency to link the distinction between country and city life with a neo-evolutionary stance. Urbanisation, he noted, 'is not a single, unitary, universally similar process but assumes different forms and meanings depending upon the prevailing historic, economic, social and cultural conditions' (1973: 129). Exploring a diversity of urban contexts had led these writers to an understanding of cities as not only diverse across the world, but also as providing the scene for an incredibly wide range of different kinds of social relations and urban attitudes in any given city. It also enhanced their sense of a lack of coherence to urban life: where Western writers had been accommodating the diversity of the city into an overarching narrative of a singular urban way of life, urban anthropologists of other cities stressed the lack of coherence to the city and the multiplicity of urban social forms.

Struggling with the association of their case-study cities with 'traditional', 'tribal' or 'folk' cultures, defined as non-urban by the Chicago School, authors insisted that what they were dealing with was indeed a form of urbanism. J. Clyde Mitchell, working in what is now Zambia, wrote in a determined tone, that 'The starting point of the analysis of urbanism must be an urban system of relationships' (1968: 48). This involved much more than simply postulating an evolutionary transition from traditional or tribal cultures, as the Chicago School approach would indicate. The persistence of 'tribalism' was not discounted. For J. Clyde Mitchell urban cultures were shaped and influenced by elements of what was then called 'tribal', or ethnic, culture. But their continued influence, J. Clyde Mitchell's colleague Epstein insisted, reflected 'processes at work within the urban social system' rather than 'vestiges from a tribal past' (1958: 239). The elements of social life that were considered by the Chicago School to be the antithesis of urban social life came to be understood as defining features of ways of being urban through studies from other parts of the world.

From the point of view of urban theory, then, the diversity of urbanisms that have been documented in different cities ought to have dislodged earlier theorisations of urbanism based only on Western contexts. As Ray Pahl (1968) noted of this body of urban anthropological research, it presented a significant challenge to urban theory, offering the possibility of provocative new accounts of urbanism and urban theory: 'This stimulus may be as stimulating and productive to research workers in the coming decade as the Chicago School was to an earlier generation of scholars' (1968: 30). And furthermore, on reviewing the anthropological literature on urbanism in different parts of the world, he concludes with the suggestion that 'there is now an urgent need for an overall understanding of urbanism in all its diversity' (1968: 39).

However, the Chicago School literature they were criticising remains very influential in Western urban theory today and is routinely included in theoretical discussions and readers. But there are very few (if any) acknowledgements of this enormous body of comparative anthropological work, which stretched over some three decades in the middle of the twentieth century. It seems their concerns and achievements have been lost to contemporary scholars of the city although, in so far as urban studies has a strong and continuing interest in urban communities, neighbourhoods and networks, for example, their influence may be forgotten but not incidental. The potential of this body of comparative work to help us reimagine what cityness entails and to reconfigure the meaning of urban modernity remains unexplored. Chapter 2 therefore stages a dialectical encounter between the Chicago School theorists and urban anthropology, hoping, in Benjaminian vein, to provoke a reconfiguration of urban theory, one in which urban modernity can be reimagined for a cosmopolitan world of cities.

CONCLUSION

> Contrary to the way in which 'non-Western' societies have been described in the scholarly and popular literatures of the West, these societies were never 'closed', 'traditional', or unchanging. Nor were they founded simply on kinship, communalism, ascriptive status, patriarchy, or any other such 'principles'. They tended rather to be complex, fluid social worlds, caught up on their own intricate dynamics and internal dialectics, the workings of which had a direct effect on the terms of the colonial encounter.
>
> (Comaroff and Comaroff 1997: 27)

One of the strongest effects of Western urban theory's ongoing association with a restricted conceptualisation of urban modernity is that it postulates a privileged link between modernity and certain kinds of societies, and between modernity and certain cities. It has done this by displacing other aspects of social life onto distant people and places. Quite wrongly, as the Comaroffs insist, these writing have portrayed rural and non-Western societies as static and closed, the antithesis of urban modernity. These assumptions have led to the identification of only certain ways of life as exemplary of cities. But they have also had profound implications for understandings of where urban cultural innovation and urban economic dynamism are located, for example.

Cities outside the West were (and continue to be) thought of as having a very troubled relation to the modern. The differences of these cities from cities in the West were elided into implications of backwardness linked to assumptions about their closer and continuing association with 'tradition', with rural places and with the persistence of supposedly non-urban practices. The locatedness of the original concept of modernity and the hegemonic position of Western urban experiences in framing intellectual fantasies of city life has left cities in poorer countries, in former colonies, or in areas outside of Western culture, to be apprehended through a static, non-dialectical lens of categorisation as other (non-Western, African, Third World).

Parochial descriptions of urban ways of life have sedimented into universalising theories of cities, with disturbing consequences. The diverse urbanisms of different cities have not been allowed to transform the theoretical categories – in this case, urban modernity – through which cities are understood. As a result, the modernity of many cities remains in question, troubled by Western urban theory's close dependence on a contrast between modernity (defined in the West's own image) and its excluded others: tradition, primitivism and difference.

But Benjamin's critique of tradition as produced within modernity offers us the opportunity to expose the fiction of urban modernity as belonging to, or invented in the West. His writings are very suggestive of the potential for seeing all cities as occupying the same historical time, open to new kinds of futures, contributing to the inventive modernities of the present. But for this, our imagination of what it means to be urban and to be modern needs to be detached from its privileged location in Western cities. A post-colonial account of urban modernity needs to be open to the intellectual resources and diverse experiences of a world of cities.

In this spirit the following chapter adopts a critical tactic of comparativism. This has the effect of dislocating accounts of urban modernity from their Western reference points and opening them up to the diverse experiences of different cities. Chapter 2 works through a vibrant debate within urban theory from the mid-twentieth century, a debate that framed much of the work of comparative urbanism at that time: the critique of Louis Wirth's account of an alienating, individualising urban way of life. Out of this encounter I hope to activate some of the cosmopolitan resources that we need for a post-colonial account of urban modernity, one that can learn from the diverse tactics of urban living around the world and that can move beyond parochial analyses of Western urban modernity to embrace a diversity of ways of being urban.

2 Re-imagining the city through comparative urbanism

On (not) being blasé

INTRODUCTION

For a few decades in the mid-twentieth century, a strong tradition of comparative urbanism provoked substantial discussion about diverse urban ways of life in cities across the world. This chapter will retrace some of these debates in order to explore some possible tactics for a post-colonial urban studies, especially the potential offered by undertaking comparative studies of city life. In pursuing this it will be important to find new ways of dealing with differences amongst cities, rather than aligning some cities with modernity and leaving others troubled by tradition, or seeing some as occupying a higher ranking because they seem more developed than others. Instead, a post-colonial urban studies will assume that there is potential for learning from the experiences and accounts of urban life in even quite different cities. In a world of ordinary cities, difference can be gathered as diversity, rather than as hierarchical ordering or incommensurability, but also without any suggestion that a universal theory of urbanism is possible.

The debates that ensued from urban anthropological research between the 1940s and the late 1960s addressed both these issues. Different cities were brought into the same theoretical debates, and efforts to establish a universal account of urbanism on the basis of the experiences of only some cities were strongly contested. The key line of debate was established as a number of anthropologists engaged with the Chicago School account of city life through close comparative research and debate (see for example, Hausner and Schnore 1965, Pahl 1968). However, by the early 1970s, the growing strength of a discourse of development meant that a sense of the differences between cities in the West and 'elsewhere' – especially what had been 'colonial' cities – hardened. The field of urban studies came to be divided by the hierarchical categorisation of different kinds of cities as developed or undeveloped (or developing or underdeveloped, depending on your theoretical orientation). This divide continues to form the basis for urban studies to this day, in which different kinds of cities are broadly thought to be incommensurable.

Before this divide became entrenched, though, writers on cities around the world contributed to developing common theoretical accounts of urban ways of life. This chapter turns to explore this tradition of comparative urbanism, lost to urban studies in the wake of developmentalism. My ambition is to mobilise the resources of comparative urbanism towards a post-colonial account which sees all cities as ordinary, rather than as defined by a-priori categories.[1] This will require us to trace theoretical and empirical paths across different kinds of cities, currently kept apart by the intellectually divisive effects of the concepts of modernity and development. There are very few precedents for this within contemporary urban studies. As a result, this chapter reaches back to an earlier moment of intense comparative work within the field, to consider some of the theoretical possibilities, as well as tactics and pitfalls, of comparative urbanism.

In a Benjaminian vein, this chapter is staged as a 'dialectics at a standstill', exploring the interaction between Western theories of urban modernity and accounts of emerging, dynamic urban cultures in African cities. These new urbanisms were richly imbricated with social relations which had been thought of as traditional and demanded a reconsideration of what it meant to be modern in the city. To stage this 'dialectic', the chapter sets out two quite different readings of urban modernity: one established through mid-twentieth-century research on the Copperbelt in central Africa (today's Zambia); the other primarily based on experiences of late-nineteenth- and early twentieth-century European and American cities. These are somewhat boldly contrasted as urbanisms of interaction and alienation in the first two sections below. The third section revisits the experiences of Western cities through these accounts of African urbanisms, and questions assessments of European urban ways of life as dominated by indifference. Finally, based on this dialectical experiment and drawing on a wider range of comparative writings, I review the potential for comparative urbanism to contribute to a post-colonial urban studies and suggest that comparative methodologies offer us a preliminary tactic for developing a post-colonial urban theory.

CITIES OF FEAR AND ANXIETY: ALIENATED URBANISMS

> The contacts of the city may indeed be face to face, but they are nevertheless impersonal, superficial, transitory, and segmental. The reserve, the indifference, and the blasé outlook which urbanites manifest in their relationships may be regarded as devices for immunizing themselves against the personal claims and expectations of others. The superficiality, the anonymity, and the transitory character of urban social relations make intelligible, also, the sophistication and the rationality generally ascribed to city dwellers.
>
> (Wirth 1964: 71)

Louis Wirth set in train almost a half century of urban research with the paper from which this quotation is taken, 'Urbanism as a Way of Life'. As

Manuel Castells noted, in 1977, of Wirth's paper: 'Its echoes, 33 years later, still dominate discussion' (1977: 77). Wirth's purpose was to relate the central characteristics of cities to the way of life followed by urbanites. He framed his paper in a very general way – as if to explain cities everywhere. And this was indeed Louis Wirth's ambition. He hoped for a 'unified body of general theoretical knowledge of urban sociology' (1964: 80) and suggested that:

> In formulating a definition of the city it is necessary to exercise caution in order to avoid identifying urbanism as a way of life with any specific locally or historically conditioned cultural influences which, though they may significantly affect the specific character of the community, are not the essential determinants of its character as a city.
>
> (1964: 66)

However, the difficulty of developing a general account of urbanism – placing all cities within a single theoretical field – is well illustrated by the evident parochialism of Wirth's theoretical analysis. As Ulf Hannerz notes, the Chicago School accounts drew primarily on research undertaken in Chicago and expressed something of the specificity of that place: 'The Chicago Studies are quite clearly set in a particular territory' (1980: 57), one in which the city population was growing feverishly and in which people were drawn from 'many countries on several continents, (resulting in) a metropolis which could seem almost a world to itself [. . .] It was hardly like all other towns' (1980: 74). Urbanism here was also closely intertwined with industrialism and capitalism (1980: 74) while many other cities had more slowly growing populations, more structured and limited forms of heterogeneity and, perhaps, less of an association (or none at all) with industrialisation or capitalism. The parochialism of Wirth's argument invited criticisms from all around the world, which we will explore in the following sections. But first, let us briefly review (for they have been much discussed in the literature) how Wirth, and the two urbanists who he drew on so heavily for inspiration, Georg Simmel and Robert Park, imagined city life in the early decades of the twentieth century, drawing as they did on their experiences of Berlin and Paris in Europe and Chicago in the USA.

The three decisive factors that Wirth identified as shaping city life were: size (number of people), heterogeneity and density.[2] Of these, heterogeneity suggested to Wirth that with the wide range of different people gathered together in large cities – to perform different functions or drawn to the city from many different places – there would be a demise of close bonds amongst people, especially ones dependent upon kinship, neighbourliness and common traditions. Instead, as we noted in Chapter 1, for Wirth, the characteristic relations of the urbanite came to be concerned with utility and efficiency and were highly segmented according to the different spheres of activity in which city dwellers were involved. These attributes, and the rational forms of behaviour they implied, Wirth considered crucial to the successful organisation of economic life in cities. However, the consequent differentiation of urban society produced a city that 'comes to resemble a mosaic of social worlds', characterised by competition, aggrandizement and mutual exploitation (1964: 74). Depersonalisation and segmentation mean that people interact in ways that are superficial and categorising, and when associations do occur, they tend to be limited to one area of a person's life, with no cumulative effect. The dense 'multiplex' social relations characteristic of 'community' life or 'folk' traditions were thought unlikely to emerge in this context. Instead, people get used to superficiality, insecurity and instability as the way of social life in cities.

Central to Wirth's account of the distant and individual nature of social relations is Georg Simmel's explanation, some decades earlier, for what he described as the 'blasé attitude' of urban dwellers. Simmel was concerned that, faced with the heterogeneity and density (to use Wirth's terms) of city life, people 'would fall into an unthinkable mental condition' (cited in Wirth 1964: 71) if they were to try and respond equally and intensely to all the people and events they observed. The concentration of people and things in the city, Simmel noted, 'stimulates the nerves to its highest achievement' (1997: 179). In response, city dwellers – well, not the stupid ones, he observes (1997: 178) – develop ways of reacting to the city with their heads, rather than their hearts. It is here that some elements of the 'rationality' of city life are generated. For the unconscious, with all its irrational, emotional depth is not the appropriate organ for confronting the often shocking and diverse phenomena of the city. Instead, intellectuality and the conscious mind protect the individual from the consequences of being in the city and keep its emotional effects remote from the depths of their personality (Allen, forthcoming). Intellectuality, then, 'preserves subjective life' (Simmel 1997: 176). Simmel notes that 'There is perhaps no psychic phenomenon which is so unconditionally reserved to the city as the blasé outlook' (1971: 329).

While this form of behaviour parallels the dynamics of the money economy, which 'hollows out the core of things' (Simmel 1997: 178) and similarly produces rational and objective, anonymous social relations, the dynamics at work in producing the blasé attitude are quite specific. They rest on the proposition that overstimulation is dangerous for the 'nerves' and that people should find ways to shut out many of the sights and sounds of the city. The blasé attitude can be likened to the condition of 'neurasthenia',[3] a popular concern of people in this part of the world, especially after the First World War. It described a form of response to psychic trauma, where people shut down their responses to the world and withdraw into themselves.

Over the years, Louis Wirth's account of urbanism as a way of life has drawn substantial criticism, as empirical researchers found evidence of very different urban ways of life and also objected to some of the assumptions that underpinned his analysis. For example, Amitai Etzioni (1959) criticises the assumption that the dense, multiplex social relations associated with urban ghettos (and which Wirth himself famously studied) will dissipate with time and converge with the alienated, indifferent relationships Wirth considered characteristic of cities. He suggests that 'non-ecological' (that is, spatially dispersed) communities amongst different ethnic, religious or language groups, for example, will persist despite suburbanisation and spatial integration. Similarly, suggestions that urban ways of life were associated with specific spatial forms (for example, suburbs or dense inner cities) was strongly disputed by Herbert Gans (1962, 1995). So a number of critical responses to Wirth emerged within the study of 'Western' cities, objecting to his characterisation of ways of life in these places.

But a very important line of critique of Wirth's paper emerged from people studying and writing about cities outside of the USA or Europe. Much of the argument here contested the suggestion that there was one urban way of life, characteristic of all cities. And many of the observations hinged on the persistence and transformation of traditional practices in the city, and the emergence of quite distinctive, but still specifically urban social relations and cultural dynamics. Urban ways of life, then, were diverse and, it seemed, constantly being invented in different cities. One of the important seams of this critique emerged from a group of anthropologists who worked on and lived in southern Africa, especially in what is today Zambia and, most specifically, the rapidly urbanising mining towns of the Copperbelt in northern Zambia through the 1940s to 1960s. The following section explores their writings on urban ways of life in these contexts with a view to establishing the post-colonial potential of comparative urban studies in which it might be taken for granted that conversations across the diverse experiences of cities such as Chicago, Paris and Lusaka would be both possible and productive.

CITIES AS SITES OF SOCIABILITY: MAKING CONNECTIONS ON THE COPPERBELT

It has been suggested that despite their apparent differences, there were many resonances between processes of urbanisation in early twentieth-century Chicago and mid-twentieth-century Zambia. Ulf Hannerz (1980) comments that although they were distant from one another in both time and space, these two contexts offered an opportunity for relatively easy comparability. J. Clyde Mitchell (1987), one of the anthropologists who spent many years researching and writing about the Copperbelt, suggests that both urban contexts were socially heterogeneous and rapidly growing, dominated by industrialisation and that both threw up similar problems of negotiating cultural differences in situations of rapid social change, widespread poverty and the challenges of workplace organisation. However this may be, the anthropologists writing on Zambia, who came to be known collectively as the Manchester School (Werbner 1984) after the institutional location of some of the key figures in the group, especially Max Gluckman, were part of a much wider community of scholars who were drawn to study cities around the world as the traditional subjects of anthropology made their way to urban settlements in search of work, money or new lifestyles. In their various research environments, these researchers took every opportunity to point out that Wirth's universal theory of urbanism was far from universal in its reach, that there were many different urban ways of life.

In Ndola, a Zambian Copperbelt mining town, anthropologist A. L. Epstein hired a young African assistant, Chandra, to work with him and asked him to prepare a record of his day-to-day experiences in that town during his research there in the early 1950s.[4] An administrative centre for the thriving Copperbelt towns of Zambia,[5] Ndola attracted African people from many different parts of the region and indeed sub-continent who passed through or settled there after the Second World War. Epstein (1969) records that the town had grown extremely rapidly in the decade after the war, and that the African population (some 50,000 people at the time he was writing) was 'ethnically mixed, [. . . with] a concomitant diversity of culture which is expressed in the wide range of languages spoken, in the distinctive and sometimes exotic modes of dress of different tribal groups, and in differences in manner and behaviour' (1969: 79). The African population was highly mobile as well as very diverse, and this made for a fluid, dynamic and very creative form of urban culture.

Magubane (1968) points out that the insecurity and mobility of the African population in Zambia, as well as many of the cultural innovations associated with this unstable urban existence that Epstein and others recorded, were a result of colonial domination and racist capitalist exploitation. The mobility of the African population was partly a result of continuing strong attachments to home places in rural areas that saw the emergence of patterns of circular and return migration. But they were also because all African people had to find employment in towns if they were to be able to access accommodation there or be given the right to remain. The taxation requirements of the colonial government pressed many Africans into migrating to towns in search of cash income, and restrictive legislation concerning the right to remain in urban areas ensured that rural areas remained important places where rights to land and accommodation could be exercised. The gendered nature of these patterns of mobility were striking, as legislation and cultural practices combined to keep many women tied to rural areas and made access to housing in cities dependent on marriage, confining many independent women to informal squatter areas and ensuring that there were many more men than women in cities during the colonial period (Hansen 1997). More long-standing patterns of mobility were associated with shifting cultivation and the combinations of patrilineal and matrilineal cultural practices in different parts of the country (Myers 2003). Nonetheless, the energy and creativity of African ways of life in these towns at this time were remarkable to the anthropologists who spent time there.

Epstein's researcher Chandra made many astute observations about the ways in which African people in Ndola built up new kinds of relationships both within and across kin groups, and enabled Epstein to develop an account of the intense sociability of African life there. Faced with the economic and social insecurity of urban life under colonial rule and given the heterogeneity of language and customary groups in the city, African urban dwellers responded with energetic initiatives to make connections with other urban dwellers. Perhaps easiest was to find and build links with people you knew from home areas, or who you knew were related to you. But the search for recognition stretched these boundaries to even remote kin members, through both male and female lines as well as to members of broader ethnic groupings and also to different, closely related or proximate groupings. As Chandra's account of his daily life shows very well, meeting new people in public and familial settings initiated a search for lines of connection. Familiar patterns of behaviour in relation to kin were extended to distant or non-kin members – what Harries-Jones called 'instrumental kin' (1975: 70) – incorporating a range of 'reciprocal responsibilities, obligations and privileges' into day-to-day relationships (Epstein 1969: 98). These relations fed in to patterns of joining political organisations – for example, the formation and organisation of the United National Independence Party (UNIP) in local neighbourhoods (Harries-Jones 1975) – and general socialising, help with finding a job or support in moving from one town to another.

If these practices were in any way traditional or tribally-based, they were most definitely, to follow Epstein, also very much a product of the urban social system (1958: 239). It was the bringing together of people from a very wide range of cultural backgrounds in the difficult conditions of colonial urbanisation which fostered a sociability that reconfigured 'traditional' practices of association in the context of the city. The heterogeneity and diversity of Zambian cities played an important part in reconfiguring 'tribalism'. Gluckman (1961: 75) notes that 'tribalism in the town operates as a primary mode of classifying the heterogeneous masses of people, whom a man meets, into manageable categories'.[6] With few clear-cut cultural boundaries amongst the many different tribal groups in the region, Mitchell noted that 'cultures tend to merge imperceptibly into one another' and in the cities the tendency was 'to reduce the wide diversity of tribes to a few categories' which were then important bases for forming associations and categorisations (Mitchell 1956: 28). 'The urban need to categorize people', Hannerz (1980: 135) summarises, 'was what "tribalism" was about in the Copperbelt town'. So what was supposedly the most modern, anonymous and urban of social relationships, the casual, fleeting and passing 'traffic' relationship (Hannerz 1980) was a site of 'tribal' identification.

The importance of 'tribalism' in the city was not portrayed in any simplistic sense. As Gluckman had already noted, 'the starting point of our analysis of tribalism in the towns is not that it is manifested by tribesmen, but that it is manifested by townsmen' (1961: 68). Far from pursuing an anthropology in thrall to the tribe, Gluckman and his colleagues considered that an overemphasis on tribalism or ethnicity would be neglectful of the wide range of urban social relationships, most notably employment relations and trade unionism, in which many African urban dwellers participated. But in the face of the distancing of urbanism from 'traditional' or folk ways of life that characterised urban theories at the time, these writers insisted that tradition was a crucial component of a very urban modernity.

It was J. Clyde Mitchell (1956) who tangled most directly with the 'paradox' of the coexistence of modernity and tradition in African cities, as he considered modern urban tribalism in his classic account of a 'kalela dance'. Celebrating a specific tribal group, with a tribally determined membership, the dance groups he observed performing in an open space in the dense municipal African township in Luanshya appropriated the style of elite urban African dress. They sang not only about their own home groups and their desire to return home, but also about their broader affiliations to closely related tribes. They performed a reconfigured relationship with tribal groups they were customarily considered distant from, or even hostile towards. To these formerly 'hostile' groups, Mitchell argued that African people now engaged in 'joking' relationships, expressing this hostility in joking, 'lampooning' ways in the songs, but apparently without animosity. The example below, which Mitchell presents alongside several other stanzas and songs that he heard, indicates something of the work that the songs did in placing a particular group within the wider, diverse social context of the city:

> You mothers who speak Tonga,
> you who speak Soli, mothers,
> Teach me Lenje.[7]
> How shall I go and sing?
> This song I am going to dance in the Lenje country,
> I do not know how I am going to speak Lenje
> Soli I do not know,
> Tonga I do not know,
> Lozi I do not know,
> Mbwela is difficult,
> Kaonde is difficult.
> All those places I have mentioned, mothers,
> Are where I am going to dance Kalela;
> Then the dancers will return to Lamba[8] country
> [. . .]
> When I finish that work, mothers,
> I shall never stay in Lambaland,
> But I shall hasten[9] to my motherland of Chief Matipa.
>   (Mitchell 1956: 7)

All of these aspects of the kalela dance added up to expressions of a distinctively urban form of tribal affiliation and to new practices of identification, support and association within the city in which the 'tribe' now merged with the African population of the Copperbelt as a whole. Singing kalela effectively took place in 'all those places' across the country from where people had assembled together in the city. As they performed before the diverse audiences of township residents, the dancers initiated the ground for affinities and new kinds of association with their fellow city dwellers. In the city, J. Clyde Mitchell argued, any persisting tribal antagonisms had to be subordinated to cooperative and peaceful interactions in the workplace and wider social life: 'The urban experience of migrants to the Copperbelt towns involved mingling with strangers of many ethnic backgrounds, and finding ways of dealing with them' (Hannerz 1980: 134). The invention of the joking relationship amongst tribes, for example, was not a tradition but a modern urban tribalism. It was the cultural work of the city dweller (Gluckman 1961: 75).

Between town and country

Epstein, Mitchell and Gluckman all stressed the urban nature of the ethnicities that they explored; none of these were considered to be simple hangovers of rural life, destined to disappear with urbanisation. They were rather invented in the city and were a specific form of urbanism, one in which revising ethnic identity, making connections and forming associations was central. For these writers there was not a progressive dichotomy between tribalism and urban modernity, the one destined to subsume the other (although see Ferguson 1999 for a somewhat different reading of their work). Rather, tribalism and urbanism each shaped and reinvigorated and, in some very practical economic as well as personal ways, depended on the other. As these writers enjoyed to point out, following Gluckman (1961), they were part of a 'single social field', stretching from town to country, as well as across the city.

This argument was not simply a theoretical nicety: in the face of the draconian anti-urbanism of the colonial administration who, like their South African counterparts, raided houses in search of unemployed people to deport to rural areas and consigned African political rights to tribal authorities, African people – and the liberal anthropologists who recorded and learnt from some of their experiences – insisted on their right to be urban. And although, as Epstein noted, 'In a Northern Rhodesian town the African is never a full and free citizen' (1964: 93), circumscribed by the dominant European institutions that shaped work and residential life in the city, the dynamic and forceful contribution of African residents to making a distinctive form of urban modernity remains very evident.

Within the racist southern African context, where African people were officially regarded as, at best, temporary sojourners in the city, the distinctive urbanism of Zambian cities refuted the authority's insistence that Africans should remain 'tribal'. Moreover, the dynamic forms of sociability that emerged in cities belied the common assumption of the authorities (echoing the Chicago School analysis) that 'detribalised' African people in the cities were a potential source of social disorder and political danger. The insistence of writers like Gluckman that African cultural practices in the city were distinctively urban, in the context of his own South African origins and Zambian politics, were not only theoretically innovative, then, but also potentially politically radical.[10] As he most famously noted:

> this patent set of observations, as well as our theoretical orientation, should lead us to view the Africans in urban areas as acting primarily within a field whose structure is determined by the urban, industrial setting. An African townsman is a townsman, an African miner is a miner.
>
> (1961: 69)

Contra both the southern African authorities and the Chicago School theories, in their view, African city dwellers were not 'natives' but urbanites; not traditional but modern.

Urban ways of life

Instead of embracing the emerging urban social order amongst African people, the ruling elite across southern Africa implemented an edifice of segregation and administrative controls over the lives of African people. In urban areas requirements for Africans to live and socialise only in certain parts of the city created deep divisions within the social and spatial fabric. The stark segregations of colonial society forced city dwellers to negotiate across quite different social environments, as well as across the many different cultural practices in the city. Devastating for African people's personal and economic lives, this divided context brought the diversity of urban ways of life into sharp perspective.

The Copperbelt anthropologists were, as a result, attentive to the diversity of social contexts within any city and their impact upon urban ways of life. Drawing on situationalism, they attempted to make theoretical sense of this urban form. The divided cities of colonial southern Africa were one inter-dependent urban field, they suggested (Epstein 1964: 99), but a field differen-tiated by ethnicity, class, race and context, as well as by cultural practices. In different situations, it was argued, people selected different behaviours; situational selection was the explanation for people slipping between rural areas and city life, for example, and finding ways to inhabit each comfortably. 'The switch of action patterns from the rural to the urban set of objectives is as rapid as the migrant's journey to town' (Southall, in Mitchell 1968: 44). Although they acknowledged the mutual influence of 'tribal' areas on city life, Mitchell and Gluckman emphasised rather the comparatively easy switching between these different worlds that they observed. The details of their ethnographies, though, suggest that these transitions were often fraught with personal difficulties for African men and women (see, especially, Mayer 1971).

In the city itself, another very difficult set of divisions – personally damaging and dangerous as much as they were sources of economic opportunity – forced on African city dwellers a diversity of urban behaviours. The African experience of dealing with the different requirements of places of residence, sociability, work and interactions with Europeans in marketplaces inspired the anthropologist's application of situationalism to the Zambian urban context (Mitchell 1987). The inconsistencies amongst the different spheres of social interaction that made up the city were resolved, according to Epstein, by the principle of situational selection (1958: xvii). The imagination of such individual segmentation is not one that sits comfortably with contemporary accounts of identity, in which explorations of hybridity dominate, but nonetheless these researchers lead us to appreciate the complexity and diversity of social relations in the city.

The classic study here, which many researchers referred to, was that undertaken by Philip Mayer (1971) in South Africa, concerning the contrasting urbanisms adopted by different groups of ethnically Xhosa people in the city of East London. With long histories of different responses to Westernisation, according to Mayer, 'Red' and 'School' Xhosa people moving to town in the late 1950s – again mostly men – created quite divergent responses to the urban context. Committed to 'traditional' ways in the rural areas, 'Red' migrants retained an ethnically 'incapsulated' life in the city, socialising with people from their home place (*amakhaya*) and maintaining close and strong ties to rural areas, which they visited frequently. For 'School' migrants, the presence of organisations and activities common to both rural and urban areas (schools, churches, etc.) and an orientation to achieving status and class improvements in a Western idiom, meant a greater engagement with the range of associational life present in the city, although interactions with rural areas often remained important to them as well.

Drawing on Philip Mayer's research, Mitchell developed the idea of the city as a 'network of networks' (1973: 310). Individuals in the city participate in varying types of networks of social relations, involving different qualities or intensities of interaction (from very intense and intimate in relation to kinsfolk, for example, to distant and fleeting in relation to people one passes on the street). Relations between people could be multidimensional (multiplex), as in the case of Red migrants with their *amakhaya*, or single-stranded as in the case of most workplace relations. He observed that there are a range of different kinds of networks in which any given city dweller participates. The character of social relations in cities is diverse and changing, depending on the nature of the activity, the social network and the nature of the situation, or place, of interaction (Hannerz 1980).

These diverse urban ways of life are evident not only in the contrast between discreet communities (as with the 'Red' or 'School' groups amongst Xhosa migrants to east London), but within the life and daily paths of individuals in the city. Any individual is involved in a variety of different types of social relations. The analysis of networks and roles became important to this group of urban theorists. They argued that people played different roles in different networks and social situations. In new and unpredictable situations, 'the actor is able to draw on several definitions of what his role ought to be' (Mitchell 1968: 59). As Mitchell noted, 'The behaviour in any one role is not always consistent with that in another, so that inconsistency and conflict appears to be characteristic of urban African life' (1968: 58). Following Epstein, he continues: 'The existence of these relatively autonomous fields of activity has led observers to see town life [. . .] "as a kind of phantasmagoria, a succession of dim figures caught up in a myriad of diverse activities with little to give meaning or pattern to it all".' (1968: 60). The conflicted divisions, diversity and mobilities of the southern African colonial and apartheid city generated an account of urban social relations as diverse, differentiated and dynamic.

The writings of the Manchester School of sociology encouraged an appreciation of the diversity of cultural practices, urbanisms and social relations of cities. Inspired by the creative and dynamic urban cultural practices amongst some of the earlier migrants to central African cities in the middle decades of the twentieth century, they provoked a sense of city life as mobile, diverse, actively associational and concerned with making personal connections (Hannerz 1980, Pahl 1968). They associated the cities of Africa with modernity and associated urban modernity very clearly with cultural practices that had previously been considered outside the realms of urban ways of life. Rather than being 'traditional' and non-urban, urban cultural practices in African cities reflected dynamic ways of living in cities. Urban ways of life, we would have to conclude on the basis of these studies, are diverse both across and within different cities.

There are many consequences of these studies for thinking about cities everywhere – and we will return to reflect on more of these in the final section of this chapter. However, as an example of how accounts of the experiences of quite different cities can learn from each other, the following section returns to the alienated urbanisms of Wirth, Park and Simmel, this time with the urban experiences from the Zambian Copperbelt in mind. This is not to suggest that we would expect to find the same dynamics in different places, although this might happen. Young and Willmott's (1986) study of family and kinship in East London noted some affinity between community dynamics there and anthropologists' accounts from other places. More likely, though, is that after thinking about the urban experiences of the Copperbelt, we might be inspired to return to accounts of the indifferent, blasé attitude of the urban dweller with some new questions. We might be drawn to think about how even the overwhelming crowds of the big city and the alienated pedestrian life of the European street, could be understood as part of a richer seam of urban social and emotional interactions.

REVISITING PARIS, BERLIN AND CHICAGO VIA THE COPPERBELT

The big city crowd provided the primary setting for the blasé indifference to events and passers-by widely seen as characteristic of the nineteenth-century European city. In Simmel's account of the mental life of the metropolis, though, the individual city dweller's apparently blasé attitude is somewhat contradictory. On the one hand, city dwellers show a measure of reserve, making city folk appear 'cold and uncongenial to small-town folk' (1971: 331). This is the observation picked up by Chicago School theorists, Wirth and Park. But Simmel's analysis, characteristically, is rather more subtle and dialectical than this and directs our attention to the underlining of emotional dynamics that go along with the blasé attitude. With the Copperbelt in mind, we might pay closer attention to these dynamics which point us towards the moments of social interaction that coexist with apparent indifference. For the sphere of indifference, Simmel proposes, is 'not so large as might appear on the surface' (1997: 179). Thus, while the surface aspect of the city dweller might be one of indifference, the internal dynamics that produce this blasé attitude deflect us from any simplistic assessment of urbanism:

> Our minds respond, with some definite feeling, to almost every impression emanating from another person. The unconsciousness, the transitoriness and the shift of these feelings seem to raise them only into indifference. Actually this latter would be as unnatural to us as immersion into a chaos of unwished-for suggestions would be unbearable. From these two typical dangers of metropolitan life we are saved by antipathy which is the latent adumbration of actual antagonism since it brings about the sort of distanciation and deflection without which this type of life could not be carried on at all. [. . .] What appears here directly as dissociation is in reality only one of the elementary forms of socialization.
>
> (Simmel 1971: 331–2)

For Simmel, indifference is a product of a mild form of antagonism: it has 'an overtone of concealed aversion' (1971: 332). Thus the achievement of indifference requires a determined deflection of the impressions of city life, with a social consequence more difficult to gloss over or valorise than indifference – latent conflict. Freedom and individuality, which Park, Wirth and Simmel all celebrate as social consequences of the city, are, according to Simmel, hostage to the pragmatics of antipathy. Furthermore, while the indifference produced in this context underpins the 'freedom' of the modern city for the individual, it also stifles individuality. So another social consequence of indifference is that it fosters a search for individual recognition. The dialectics of individualism and individuality (Frisby 2001) play themselves out in the psychodynamics of the big city crowd: seeking recognition from the slightly averse, indifferent city dwellers in an anonymising economic and social world, some people are drawn to explore their individuality in extremes of mannerism and appearance, exaggerating their uniqueness (Simmel 1997: 183–4). The relation between the individual and society is profoundly altered in the city, according to Simmel. But exactly how the city will shape this relation, he concludes, will be determined by a struggle between the dynamics of individual independence (bound up with indifference and anonymity) and the elaboration of individuality, which depends on recognition and social interation. The city provides the 'arena for this struggle and reconciliation' (1997: 185).

The dynamics of indifference, then, are tied into intense emotional investments in only slightly submerged relations of public antagonism and is at odds with other ways of being in the city crowd, including extravagant and public performances of individuality. Rationality and indifference are in no way predetermined outcomes of city life, but rather the site (and source) of conflict and struggle. In any event, and following Simmel, the blasé attitude has far too much to do with the complex emotional experiences of city life to constitute an unambiguous sign of rationality and distance. Feminist scholars of modernity and the city in Europe would agree with this – suggesting that the position of detached observer of the city crowd was not readily available to either men or women. Women because their presence in public spaces was deeply contested although there is plenty of evidence that many women participated in city crowds, visited department stores and enjoyed the spectacle of city life (see Wolff 1985, Buck-Morss 1986). And men because, as Elizabeth Wilson puts it, 'the interpretation of the *flâneur* as masterful voyeur underplays the financial insecurity and emotional ambiguity of the role' (1992: 106). More than this, the fragmentary and overwhelming experience of city crowds, she suggests, has a disintegrating effect on the masculine identity (1992: 109). If the blasé attitude is meant to be a defence against the shocks of city life, this feminist analysis suggests that it is relatively unsuccessful and that instead the experience of the city can have strong emotional consequences for both men and women.

Benjamin, like Simmel, observes that for those uneasy about the big city crowd in nineteenth-century Paris, perhaps feeling discomfort and fear when confronted with the vast unknown masses, there were a range of tactics that could be adopted in defence. One was to appreciate and take comfort from the numerous popular stories (*feuilleton*) of city 'types'. Sketching with some imaginative detail the concerns and habits of the different kinds of people likely to be found in the crowd, the sensitive city dweller on reading these narratives could rest easy as to what their motives and lives might be like (Benjamin 1997: 37). Not unlike the simple categorisations that observers thought 'tribal' affiliation permitted on the streets of Zambian cities.

Picking up on Simmel, and on Freud and Proust, Benjamin elaborated on how adopting a blasé attitude could allow the conscious self to parry the shocks of the city, ensuring that the jarring and sometimes disturbing aspects of city life were simply noted in conscious experience, as part of the flow of life. This would prevent them from generating a strong emotional response or causing fright and, in his view, therefore not be marked on the unconscious. Failure to screen against shocks in this way, it was thought, could have a traumatic effect, causing unpleasant experiences to be recorded in the unconscious and perhaps even requiring the work of dreams and conscious recollection to quell resultant anxieties, or to work through, organise and order memories at a later time (1999c: 157–8).

But there is a downside to this strategy for indifferent urban living, Benjamin observes. It's not very good for a poet: 'If it [an incident] were incorporated directly in the registry of conscious memory, it would sterilize this incident for poetic experience' (Benjamin 1999c: 158). Benjamin's understanding of Charles Baudelaire, 'lyric poet in the age of capitalism', was that he 'placed the shock experience at the very centre of his artistic work'. Rather than dealing with the shocks of the city in a careful, distant way, Baudelaire chose to 'parry the shocks, no matter where they might come from, with his spiritual and his physical self' (1999c: 160). As a 'hero' of modernity, Baudelaire strove to fully experience its exciting and destructive consequences in order to inform the emotionally laden lyric poetry he adapted to the modern era (Benjamin 1997). In a less dramatic way, ordinary city dwellers could enjoy the attraction and allure of the crowd. Thus the diverse and changing landscape of the city could be actively sought out by city dwellers, caught up in a form of *flânerie*, or simply enjoying a stroll in the city or observing the crowd from a suitable vantage point. Not parried, the 'shocks' of the city could be actively sought to be part of the deeper and intense memories that make up the long-term experience of life.

The experience of the big city crowd shaped the phantasmagoria of European urban life. And of European and American urban theory. A site of anxiety and alienation as well as fascination and entertainment, the mass public life of the dense and heterogeneous city captured the imagination of urban theorists and urbanites alike. Although drawn to the crowd as a source of fascination, Baudelaire himself was finally defeated by the grim pragmatics and sometimes hostile nature of city life – much as Elizabeth Wilson's feminist analysis might predict. Telling a story from Baudelaire in which the 'lyric poet' loses his halo in the mud of the busy street, Benjamin observes that 'The lustre of a crowd with a motion and soul of its own, the glitter that had bedazzled the flaneur, had dimmed for him' (1999c: 189). Baudelaire's heightened attention to the excitement of the crowd and his rather brutal experiences of the rigours of city life alert us to the intense and varied emotional life of the city (Pile 2005). Behind the masks of busy indifference, boredom, idle strolling or the passively waiting person, there is, it seems, a lot going on. And even for the consciously indifferent city dweller, more intense experiences have a way of regularly breaking through the protective shield of consciousness, so that the sights and sounds of a city street draw us into relationships with those around us and penetrate to the deeper realms of our experience and unconscious memory.

These accounts of city life situate the urban dweller in a vital world of fantasy and interaction with others, rather than in an alienated world of indifference. Fear and anxiety might drive people to a superficial distancing from others, but even in this distanciation the drama of unconscious life and conscious reactions draws passers-by into forms of association and inter-action, whether of hostility or fascination. In this way, the collective nature of city life is constituted imaginatively; the sociability of urban existence is forged in the intangible relations of the street and the phantasmagorias of city life. The big city street in Europe, considered after accounts of Copper-belt urbanism, becomes a site of interaction and not simply of alienation.

URBANISM IN WORLD PERSPECTIVE

> In Africa, the age of new cities, cities that have connotations different from those generally associated with 'urbanism', appears to have begun.
>
> (Kuper 1965: 22)

Placing these two different accounts of urban ways of life next to one another as this chapter has done poses a significant challenge to the divided nature of contemporary urban studies. In a world of urban scholarship carved into separate spheres of developed and developing (or underdeveloped) cities, the fortunes of Zambian cities and the experiences of city life in Paris are widely considered irrelevant to one another.[11] And yet the efforts of compara-tive urban anthropology through the middle decades of the twentieth century routinely questioned theoretical conclusions reached in a Western context, drawing on evidence from what we might today label 'Third World' cities. Commenting on Parisian urbanism through the lens of Lusaka, for example, would have been considered most appropriate, and themes such as racialised exclusion, immigrant experiences and urban ways of life were regularly commented on across quite divergent urban contexts. Moreover, as we saw in our discussion of the Manchester School, there is no sense in which these African urbanisms could be cast outside of the realm of the urban or placed on the side of the primitive or tradition; they were intrinsically urban cultural phenomena. Following in the tracks of these writers, a post-colonial account of urban life would strive to reimagine the city through attending to a diver-sity of urban experiences. In this section, I want to suggest that comparative research, like these mid-twentieth-century studies, represents one important step towards a postcolonialised urban studies. This section sets out some of the implications of anthropological studies of different urban ways of life for wider urban theory and then it turns to consider whether comparative urban research can provide a way forward for a post-colonial urban studies.

Beyond the blasé attitude

Like the Copperbelt anthropologists and many other comparative urbanists, Oscar Lewis observed that the blasé attitude of Western urban dwellers, which Simmel, Park and Wirth had seen as a cornerstone of urban sociability, was not representative of urbanism elsewhere. He noted of his study area in Mexico City that the residents there 'showed much less of the personal anonymity and isolation of the individual which had been postulated by Wirth as characteristic of urbanism as a way of life' (1973: 130). He did not stop, though, at simply noting the differences from the US analyses. And with J. Clyde Mitchell, Epstein, Gluckman and South African anthropologist Philip Mayer, he went on to suggest that 'there are many ways of life which may coexist in the same city' (1973: 130). He insisted that his observations from Mexico should make a difference to how cities in other places are understood.

The theoretical innovations of this body of work were substantial. The conclusions reached by the Manchester School anthropologists concerning the diversity of ways of life in cities represented a significant synthesis of a wide range of research results emerging in many different contexts at the time (see also Abu-Lughod 1961, Gans 1962, Dewey 1960). Drawing on network analysis and his observations on the Copperbelt, Mitchell proposed the idea of the city as a 'network of networks' (1987: 310), in which individuals are located within varying types of networks of social relations, involving different qualities or intensities of interaction, varying from very intense and intimate in relation to kinsfolk, for example, to distant and fleeting in relation to people one passes on the street – the classic blasé urban attitude. Social relations between people in cities might be either multidimensional (multiplex – as Wirth characterised folk culture) or single-stranded, as Wirth described 'urban' sociability.

Counter to the Chicago School's suggestion that urban life is principally characterised by single-stranded, distant and often blasé interactions, Mitchell suggested that there are varied kinds of networks in which city dwellers are located and that, depending on the nature of the social network and the nature of the situation, or place, of interaction, the character of urban social relations is diverse and changing. These differences are apparent not only between discreet communities, as with the 'Red' or 'School' groups amongst Xhosa migrants to East London (Mayer 1971). Within the life and daily paths of individuals in the city a variety of different types of social relations are also evident. Both Lewis and Mitchell complained that the Western urban sociologists treated the city as one uniform social system – whereas their work had led them to be very conscious of the multiplicity of city life. Through this comparative analysis we can observe some of the first lineaments of what might be an account of 'ordinary cities', which draws inspiration from a range of different cities and which places all cities within the same temporal and analytical field while appreciating diversity across and within cities.

For writers such as those exploring Copperbelt urbanisms in the mid-twentieth century, local debates and political concerns were important, as we saw, but contributing to urban theorising at the broadest level was also a strong motivation for their work. Far from resting easy with the 'difference' of these cities – as colonial, African, or 'traditional' – they insisted that 'The comparative background for these analyses is urban sociology in general' (Gluckman 1961: 80). Drawing perhaps a hyperbolic picture to advance his case, Gluckman continued to suggest that 'In all these respects, Central African towns differ only in degree from any town, anywhere in the world probably' (1961: 79). Far from being distinctive to central Africa, or colonised contexts, tribalism, he argued, characterised cities everywhere:

> Tribalism acts, though not as strongly, in British towns: for in these Scots and Welsh and Irish, French, Jews, Lebanese, Africans, have their own associations, and their domestic life is ruled by their own national customs. But all may unite in political parties and in trade unions or employers' federations. Tribalism in the Central African towns is, in sharper form, the tribalism of all towns.

> (1961: 76)

The enthusiasm for comparativism drew observers to notice similarities – like the 'tribalism' of British society – and differences across a range of contexts. Mitchell (1987) writes that 'In principle what is being achieved in comparative analysis is that the manifestation of certain regular relationships among selected theoretically significant features in the two instances is being demonstrated by showing how the operation of contextual variations enhances or suppresses the expected pattern' (1987: 244). The ambition was to understand the nature of social life and interaction in cities, which it was assumed would vary with different structural contexts (racial orders, rate and nature of economic growth, political power) and also across different situations and contexts in the same city, or even in one person's trajectory through the city.

Within a comparative framework, investigations in the USA and in Africa could inform one another in the task of understanding social processes in these cities (Mitchell 1987: 245). Mitchell does this, for example, by showing how his analysis of migration to cities in Africa could enhance the understanding of Chinese immigrants' experiences in US cities (1987: 292). With a similar background context of racialised discrimination and persistent cultural traditions, the segregation and relative incapsulation of Chinese migrants in US cities bore important similarities to African urban migrants in southern Africa.

While the critique of the 'blasé' attitude of urban dwellers emerged from writers concerned with different forms of urban sociability in cities outside the USA, by the late 1960s this critique had come to shape analyses of cities in many parts of the world. Ray Pahl noted, for example, that there was now an 'overwhelming body of evidence that central areas of cities differ from what Wirth and many others have suggested' (Pahl 1968: 267). His citations for this stretch from London, Boston, Delhi, Cairo, East London, Lagos and Sumatra to Mexico City. Collectively these studies conclude that 'urban villages exist in the centre of cities in which there is a high level of social cohesion based on interwoven kinship networks and a high level of primary contact with familiar faces' (Pahl 1968: 267), and together they add 'greater depth to what now seem the somewhat superficial formulations of Wirth and Gans' (Pahl 1968: 279).

Dealing differently with difference

Anthropologists, such as those of the Manchester School, show us that urbanism as a range of distinctive and vibrant cultural experiences in diverse cities is not easily captured by an urban sociology embedded in Chicago, Berlin and Paris. This remains a problem for contemporary writers: in the late 1990s Karen Hansen prefaced her account of urbanism in Lusaka by pointing out the need to 'describe urbanism in Zambia in local terms rather than through Western referential assumptions about what the city ought to be like' (Hansen 1997: 7). And Ellen Hertz (2001) raises the relative irrelevance of the Chicago School studies to contemporary Chinese urbanisms, since anonymous urban interactions have drawn far less attention there than accounts of the inventive ways in which social networks and associations are sustained and created in cities. As she notes, 'The result of this paradigm of sociality is a focus on embedded, long-term networks of reciprocity and competition in structuring urban Chinese society' (Hertz 2001: 277). She suggests that a diversity of forms of social relation – 'multiple codes of belonging and distancing' (2001: 285) – need to be attended to. But she also argues that what it means to be part of a crowd, to be 'faceless' or anonymous, can be very different across Chinese and Western contexts.

For comparative urbanists, exploring the diversity of ways of urban life was a valuable research methodology to test the claims of the Chicago School, for example. The anthropologists I have been referring to all felt strongly that their work in places such as Lagos, Lusaka, Mexico City or Djakarta could feed directly into urban theory and could in fact also improve and extend understandings of Western cities and of cities everywhere. Certainly, their work offered the resources to contest the exclusion of certain cultural practices (held under the sign of the primitive) from understandings of cityness, and as we have seen in this chapter can be used productively to reimagine theoretical accounts of urban life.

But for all its benefits, it can be all too easy to snap comparative accounts of different cities back into the universalising ambitions of Western theory. Amongst his many good reasons for pursuing comparative urban research, Gideon Sjoberg noted that: 'Urban ecological and social structure in America cannot be understood without recourse to comparative sociology.

Only through a comparative approach can we separate the general from the particular' (Sjoberg 1959: 359). Such an approach would ensure a 'sound body of knowledge' (Sjoberg 1959: 338) and enable American scholars to tell which aspects of their cities were due to urban-industrial life common to many other cities and which to their unique social and cultural dynamics (1959: 349). Of course, there are strong reasons to imagine that knowing about other cities might help one understand one's own context better. In a world of interconnected cities, understanding different forms of urbanism could be important to making sense of the diverse ways of living, brought together in most heterogeneous urban contexts. A comparative urbanism *for* America,[12] though, might be contrasted with a more abstract and less obviously self-interested search for the universal dynamics of city life that, we have seen, was advocated by Louis Wirth.

However, and as we have noted, Wirth's attempts at establishing a universal theory of urbanism ran aground on the charge of parochialism. One response to the patently parochial nature of universal claims was to turn this ethno-centrism into a strength (Hannerz 1980: 75). Perhaps, Redfield and Singer (1954) proposed, each cultural tradition shapes its own unique type of city. That different cultural regions and different forms of historical economic and social structure generated different kinds of urbanism became an important theme of later comparative urbanism (Sjoberg 1959, Wheatly 1967). Indeed, there is an important insight to be gained here, which is that wider structural and economic forces, including the dynamics of imperialism, for example, or trade and international finance regimes in the contemporary period, might play an important role in placing some cities in a common relationship to external actors and wider processes. The attempt to identify a category of colonial city or a 'post-colonial' city, signifying the shared experiences of colonial rule and independence from colonialism, would be relevant here (Jacobs 1996, Kusno 2000).

Inadvertently sounding the death knell of comparative urbanism though, these 'structural' distinctions were consolidated in 1970s urban theories into assessments of a world of cities divided by uneven development (Santos 1979, Friedmann and Wulff 1976). For Hannerz, the differentiation of urbanism on the basis of structural, economic or cultural features had only one certain outcome: a parochial and divided form of urban theory (1980: 76).

So, the potential of comparative urban research to enable a post-colonial urban studies goes along with some possible pitfalls. Ethnocentric reasons for comparison, the often parochial nature of abstract theorising and claims that different categories of cities are incommensurable are all potential dead ends. Any path towards a post-colonial urban theory, then, will have to negotiate the twin dangers of proposing particular experiences as universal theory and demarcating incommensurable domains of urbanism. Post-colonial urbanism will need to track across the differences amongst cities to build a theoretical account of city life without assuming its universal applicability or segmenting the world of cities into categories.

These difficulties have dogged urbanists over the last century, albeit in different ways at different times. In their search to refuse the incommensurability of the urban experience in central Africa, the Manchester School of urban anthropologists insisted on the potential relevance of their findings for very diverse kinds of cities. By contrast, the Brazilian geographer Milton Santos was eager to slough off the domination of Western-centric urban studies and motivated for a reorienting of urban studies so that it would be more conducive to understanding the 'reality and internal dynamic of urbanisation in underdeveloped countries' (1979: 4). His complaint was that many of the comparative efforts of urbanists through the 1950s and 1960s simply transferred terminology from the environments with which these writers were already familiar, namely the urban experiences of Western countries. As he continued, 'The fundamental mistake that many researchers have made is to rely on comparisons between the developed world and the less developed world. In this way concepts formulated on the basis of data from developed countries have been indiscriminately applied to Third World Countries' (Santos 1979: 6). Instead he recommended that writers concerned with Third World cities begin with the realities of these places and that, on this basis, they could 'reach very different conclusions from those researchers who depend upon spurious cross-sectoral comparisons' (Santos 1979: 6).

Together with others who have charted the specific forms of urbanisation in 'Third World countries' since the 1970s (McGee 1971, Roberts 1978), in trying to avoid the hegemony of Western urban theory Santos established a firm demarcation between the urbanism of 'advanced' Western countries and the characteristic forms of cities in underdeveloped, or developing Third World countries. In this manoeuvre the grounds for assuming the incommensurability of accounts of city life in the West and in other contexts were firmly set for the next decades.

In the wake of the increasing inappropriateness of this division between Third World urban studies and more Western-centric urban theories in a globalising world, a current generation of urban scholars are eager to imagine new ways to construct accounts of cities that stretch across different contexts, to find new ways of dealing with difference. But some recent attempts to explore cities in a comparative way have struggled to escape from categorising assumptions about the differences amongst cities. Dick and Rimmer (1998), for example, proposed an assessment of whether 'Third World' and 'Western' cities are becoming more like each other over time (or not).[13] Writing to the title of 'Beyond the Third World City' they suggest that there have been periods in which Third World and Western cities have converged (such as from the 1880s to the 1930s when there was an increase in economic and political control exerted by metropolitan powers through colonial rule, trade, investment and new transport technologies) and periods when they have diverged (such as from the 1940s to the 1970s, with the breakdown of colonial political and economic control, the rise of indigenous administration, the disintegration of infrastructure and prevalence of the informal economy). More recently, they note that 'Convergence between urban forms in metropolitan countries and Southeast Asia was renewed in the 1980s by increasing trade and investment and the application of telecommunications and high-speed transport' (1998: 2306).

In their view, south-east Asian cities have seen First World forms of investment, such as large-scale private land development and the proliferation of shopping malls – 'clearly First World not Third World' (1998: 2316). These developments are dominated by American architectural influences as well as by another set of American concerns, the 'perceived deterioration in personal security' with crime, racism and a sense that 'public space has become an area of uncertainty' (1998: 2317). Alongside shopping malls, gated communities have also appeared on the landscape. They are clear that the processes underpinning these changes in the urban landscape are somewhat different in south-east Asia than in the USA. But they conclude that 'The emerging urban forms take after North American patterns to a remarkable degree that has yet to be recognised, let alone explained. [. . .] Scholars need to challenge prejudices which have allowed them to partition the world into separate spheres according to their own particular areas of expertise' (1998: 2319–20).

This is certainly a sentiment that I share and one that goes some way to addressing my concerns about the divided nature of urban studies. Their strategy, though, still depends on the categorisation of cities (Western and Third World) and encourages us to assess one category of cities in terms of another; even if the purpose is to show that the categorisations may be more muddled than originally thought. For Dick and Rimmer, the comparability of these sets of cities relies on identifying common features across them, or tracing the transnational interactions that shape life in both kinds of cities.

This is a common approach in recent comparative writing on cities: identifying convergences or tracking similar processes across different cities. So as new kinds of convergences become apparent – such as the spread of service-sector firms across the globe – a wider range of different cities has been drawn into comparative analyses. There is an implicit suggestion, then, that growing convergences between cities of the 'North' and 'South' (see, for example, Cohen 1997, Dick and Rimmer 1998) make them more comparable. In my view this is more than a little misleading – must we wait for social or spatial phenomena to become the same before we can learn from experiences in different kinds of places? The experiences of the Copperbelt anthropologists suggest not. For the broadest conclusion of their work is that accounts of urbanism can stretch across quite different kinds of cities while appreciating the diversity of urban ways of life. This is a crucial contribution to our search for grounds for a post-colonial urban theory. As Ulf Hannerz notes: 'World urbanism thus exhibits many variations and exceptions, few universals or regularities: Brave attempts, such as Wirth's, to formulate a common urban pattern have in the end supplied strawmen rather than lasting paradigms' (1980: 98).

He goes on to suggest that we need to 'recognise the variety of urbanisms and their internal differentiation' (1980: 110). Thinking across Chicago and the Copperbelt, or learning from ways of urban life in different contexts, was an intellectual challenge that an earlier generation of urban scholars took very seriously. In the current era of globalising connections and increasing links between quite different kinds of cities this challenge once again confronts the world of urban studies.

CONCLUSION

The examples of comparative urbanism reviewed in this chapter invite us to reconfigure the grounds and the tactics for urban studies across different contexts. They insist that we can place different kinds of cities within the same field of urban theory without suggesting that all cities are alike; and without the need to divide cities around the world into groups sharing common features and distinguished from other categories of cities. The comparativism of an earlier period, which was happy to bring together New York and Lagos, or Mexico City and Madrid, should give contemporary scholars some inspiration for finding the grounds for common understandings and theoretical insights across the divides of modernity and developmentalism. Similarities and differences, I would suggest, are promiscuously distributed across cities and do not neatly follow the lines of cultural, regional or structural categorisations of the world of cities.

A post-colonial urbanism will be open to theory travelling in any direction: from Chicago to the Copperbelt, but also back again. Of course like all travel in a globalising world, there are obstacles to overcome, historical ties and entrenched power relations that shape the directions of influence, and limited channels to follow (Tsing 2000). And there is no a-priori reason why ideas and writing in one context have to make sense in another. But the opportunity is there, to establish conversations because of, and across, strong empirical and theoretical differences (see for example Oliviera 1996). As a tactic for post-colonial critique, a form of comparative studies that rests not on a-priori categorical and structural similarities within groups of cities but on a cosmopolitan and curious theoretical endeavour, inclusive of all kinds of cities, might stimulate and transform the divided form of urban studies.

The practices of urbanism, we learn from critiques of the Chicago School, are diverse both across and within cities. Understanding and changing life in cities needs to attend to the diverse sociabilities of city life – the many different ways of being urban and modern. For a while, European fantasies about alienated city life came to stand in for urbanism everywhere, but scholars of other contexts noticed a diversity of ways of living in cities. Bringing different accounts of city life into conversation, as we have done here, can transform our understandings of city life, in this case making us alert to the rich diversity of ways of living in cities. Whereas categories such as Western and colonial, or developed and underdeveloped, restrict our understandings of city life, insisting that all cities are best understood as ordinary will ensure that cities everywhere can inform urban theory, and also clarify for us that diverse forms of social relationships coexist in the dense social networks that come together in urban contexts. A post-colonial urban theory, I suggest, can build on comparative traditions to think through the diverse experiences both within and across different cities. Certainly, to understand ordinary cities urban theories will need to travel but, like the blasé attitude, they will never be the same again if they do.

3 Ways of being modern

Towards a cosmopolitan urban studies

INTRODUCTION

The meanings of cityness and modernity urgently need to be revitalised so that, in a post-colonial spirit, the diverse ways of life in cities around the world, the many different ways of being modern, can shape the intellectual terrain upon which cities are imagined. A post-colonial urbanism should find many resources in the current era of globalisation. Cities in so many parts of the world are routinely tied together through translocal processes, including urban design, development, processes of twinning, a competitive search for high international profiles, movements of people and resources, the circulation of information and ideas. The need for a more cosmopolitan form of urban studies has never been more apparent. Ordinary cities, then, exist in a world of cities linked through a wide range of circulations – of people, ideas, resources – in which cities everywhere operate both to assemble diverse activities and to create new kinds of practices. This chapter therefore introduces a new tactic for a post-colonial urban studies: a form of theorising in which the resources for thinking about cities are as cosmopolitan as the cities that are being theorised.

In Chapters 1 and 2, I set out some of the initial tactics required to initiate a post-colonial account of urban modernity. First, we established the need to dispossess the West of its privileged relation to an originary modernity. Even accounts of urban cultures that announce the presence of 'alternative' modernities (Goankar 2001), or that pluralise the experience of the modern, tend to maintain as primary a 'modern' supposedly invented in the West. It is often in reference to the apparently original modernities of the West that translations, adaptations or hybridisations of the 'modern' are recorded. I explained that I want to go further and decouple the modern from its privileged association with the West. More specifically, I want to dislocate accounts of 'urban modernity' from those few big cities where astute observers elaborated on the broader concept of 'modernity', placing it in a privileged relationship to certain forms of life in these places.

The second tactic that I proposed in reconfiguring accounts of urban modernity was to refuse to place it in relation to something called 'tradition'.

Opposing 'tradition' to 'modernity' is a flaw that we can trace through many existing accounts of the modern. Instead, I envisage a concept of the modern that appreciates that people in many different places invent new ways of urban life and are enchanted by the production and circulation of novelty, innovation and new fashions; a concept that explores the wide diversity of ways in which, in Marshall Berman's (1983: 345) phrase, people make themselves 'at home' in a changing, perhaps modernising world.

Finally, in this chapter the tactic of adopting a cosmopolitan approach to urban theory will address head on the harmful habit of viewing the embrace of novelty as 'innovative' in Western contexts, but 'imitative' in many others. Racism and developmentalism, even critical theories of underdevelopment such as that proposed by Milton Santos, have left a trace in which some cities are seen as architects of their own future while others are troubled by the sacrifice of creativity to imitation. I intend then to appropriate autonomy and creativity for all cities. Where accounts of modernity in poor places have emphasised mimicry and have seen modernity as undermined by tradition, by poverty and, especially, by racial difference, I want to stress the centrality of appropriation to modernity everywhere and to insist that modernity is borrowed, invented and valorised in both wealthier and poorer cities.

The first set of examples below track the circulations and inventions associated with modern architecture, in both its internationalist and popular art-deco forms. The journeys of architectural design across a world of cities chart something of popular enthusiasms for the city, for the impressive skylines produced by massed skyscrapers, for the excitement of new styles and designs for the urban environment. The chapter also considers circulations of popular cultural practices, as people remake and reimagine cities and city life in places such as New York, Rio de Janeiro, Kuala Lumpur, Lusaka and Johannesburg. The argument is that in these cultural practices, shaping both the built environment and lived aspects of city life, inventiveness is dispersed across different cities, and cities are characterised by many different ways of being modern. So the examples in the chapter bring together various aspects of modernity – built form and design as well as style, fashion and wider cultural practices – and explore the diverse ways in which people in different cities produce, experience and engage with urban modernity.

Modernities, I will suggest, are both made and appropriated in many different cities: urban modernity is a truly cosmopolitan phenomenon and can belong to any city and any people that choose to claim it. The importance of this analysis lies not only in developing better or more diverse renderings of what it means to be modern or to live in cities. As we will see in later chapters, imagining better futures for cities depends on having a strong sense of their distinctiveness and creative potential. It is my hope, then, that cities everywhere can be framed as 'modern', defined neither in hierarchical relation to others, nor in terms of their lack or deficiencies. Ordinary cities invite us to an appreciation of all cities as sites of the production and circulation of modernity.

THE COSMOPOLITAN CIRCUITS OF ARCHITECTURE: MAKING THE MODERNE/MODERN

Through the twentieth century there was little that stood more iconically for the achievements of modernity than the towering skyscrapers and the glittering skylines of rapidly growing cities. The architecture and massed form of modern cities intersected with wider currents of modern culture and with cultures of modernisation – the excitements many urban dwellers and others felt for the achievements of the 'machine age', the speed of innovations in transport and the often insistent verticality of new urban developments. In many cities around the world the fantasies of modernity were fed by the dynamic physical fabric and daily experiences of the city form itself.

But in some cities the excitements of modern urban form and design were almost drowned out by anxieties – anxieties about relying on borrowed ideas, about importing materials, about betraying local cultural forms and economic realities. In others, these very borrowings enhanced an already confident sense of originality. Take the paradox of two cities – Rio de Janeiro and New York – not so distant from each other, both part of the 'New World' of the Americas, on the same side of a continent, and both strongly associated with modernist architectural and urban developments. The skyline of New York and its intensely cosmopolitan culture have perhaps more than any come to stand in for the modernity of America and its cities in Western cultural imaginations (Ward and Zunz 1992: 4, Holleran 1996: 568). Similarly, Rio de Janeiro's wealth of modernist architecture, its visually dramatic mountainous setting, world-famous carnival, and the ambitious urban plans of the elite in the early decades of the twentieth century might be expected to add up to an icon of modernity – and at certain moments it has done so (Fraser 2000). And yet locals worried that these achievements were simply 'copies' of European originals. More than this, while Rio settled down to a place in Western urban studies as a 'Third World city' (see for example, Gugler 1997, Burgess et al. 1997, Drakakis-Smith 2000), New York has continued to play a role as an exemplary modern city – even if now the 'landscape of modernity appears very old indeed' (Ward and Zunz 1992: 14) and is increasingly dwarfed by the skylines and urbanisms of places like Hong Kong, Shanghai and Kuala Lumpur.

What is important about these divergent positionings of two culturally impressive cities is not simply that New York and Rio de Janeiro have come to occupy different places in the international circulation of iconic urbanisms. Local cultural narratives about the meaning of modern urban forms in Rio de Janeiro and New York have been quite different too. And it is these different accounts of being modern that illuminate some of the difficulties that contemporary urban studies faces in speaking across diverse urban experiences. For, like *cariocas* (residents of Rio de Janeiro), in their ambitions to be contemporary, New Yorkers appropriated styles, ideas, forms and innovations from elsewhere and even set out to imitate other places. Writers, for example, hoped to recreate the achievements and atmosphere of literary Paris in 1920s New York (Cowley 1994). But, unlike their anxious counterparts in Rio de Janeiro, New Yorkers, for much of the twentieth century were generally very confident in designating these imported, sometimes reworked forms as their own. Ann Douglas traces this to the rising economic strength of the USA and a growing confidence in the 'youthful' vitality of the New World that in 'the [post-First World War] era [. . .] saw America become the world's most powerful nation [and] also saw New York gain recognition as the world's most powerful city' (Douglas 1996: 4). It was at this moment that New York was expanding vertically, creating the skyline that was to symbolise this newfound confidence.

Ambiguous modernity in New York

Even as the most powerful city in the world, and perhaps the most popular international symbol of urban modernity, New York was intimately dependent on borrowings from other places. Writing in the early 1950s, Lewis Mumford observed that:

> One would not do justice to the American tradition in architecture if one neglected the part played in our own development by forces originating outside our country. Our very modern means of communication and transportation have made us conscious, as no other age before, of the existence of other regional traditions and cultures.
>
> (1972: 27)

He notes amongst other influences the importance of the simple, clean forms of Japanese design for the work of Frank Lloyd Wright, the importation of the verandah from India, the dispersed international origins of the industrial machines that inspired architectural innovations from the late nineteenth century on, as well as the heavy European *beaux-arts* influence on American architectural design. Mumford measures the achievements of nineteenth-century American architecture and engineering against the advancements made in the then more economically powerful United Kingdom: 'our country lagged behind the superb constructions of Rennie and Telford and Brunel [but by] . . . the third quarter of the nineteenth century we had almost come abreast of the British' (Mumford 1972: 19). With the invention of a new process for making steel in Britain, he observes, 'The engineers and architects of Chicago suddenly leaped ahead' (1972: 19). Although betraying his desire to see America overtake the Old World on which it had depended for so long, he nonetheless concludes that 'We are most deeply ourselves, we Americans who have come together from every other land, when we are most actively part of a world-wide community' (1972: 30).

This interplay between being American and being intensely cosmopolitan framed understandings of New York: as the publishers of a publicly funded initiative, the Federal Writers' Project, put it soon after the 1930s depression, New York was 'the most cosmopolitan yet at the same time the most American of all cities' (Federal Writers' Project 1939: v). Introducing this collection, the wide range of connections shaping this city fired the imagination of author Susan Ertz:

> There, on her narrow and populous island, New York sits facing the Atlantic, knitting up threads from 48 different states together with threads from London, Singapore, Moscow, Berlin, Paris, Buenos Aires, Honolulu – threads from every corner of the earth; they a part of her, she a part of them, all bound together in the vibrating, sensitive, indivisible fabric of the present day.
>
> (Federal Writers' Project 1939: 19)

Even as architectural achievements have come to symbolise the modernity of American cities, especially in New York, Mumford's careful account of the diverse origins of what are often portrayed as Western innovations is crucial. It helps us to 'provincialise' Western urbanism – to place it as just another cosmopolitan urbanism, alongside a range of different ways of being urban and modern. Helpfully, too, Mumford and others observe that the self-confidence that enabled the diverse origins of Western urbanism to be forgotten was only relatively recently acquired in the USA (see, for example, Douglas 1996: 184, Federal Writers' Project 1939: 181). During the nineteenth century the dominance of architectural designs and urban planning which aimed to reproduce the best of the classical and *beaux-arts* designs from Europe in America, produced an anxiety about whether they were being American. But the United States emerged from the First World War in a much stronger economic position than its European rivals; with this came the possibility of a more confident approach to urban innovation.

The inferiority complex, which Mumford recalls characterising architectural criticism through the late nineteenth and the early decades of the twentieth century (1972: ix), was increasingly replaced by a strident and avowedly indigenous urbanism that placed New York at the centre of twentieth-century modernity – at least as far as New Yorkers were concerned. The skyscraper especially was hailed as a specifically American invention and, indeed, the form became quintessentially associated with New York, despite the fact that the original breakthrough in vertical construction had been made in Chicago (Taylor 1992). And Chicago was only one of many different influences on New York's iconic skyscrapers. When art-deco design travelled to New York from the Paris exhibition in 1925, it intersected with the dynamics of zoning legislation and the economics of commercial office space and elevator costs to shape the form and decoration of a number of the buildings that still dominate the distinctive skyline of this city (Willis 1992).

The stepped-back form of skyscrapers such as the Rockefeller Centre, the Empire State Building and the Chrysler Building fed back into the symbolism of 1930s Western modernity, in turn becoming icons of the art-deco movement (Benton 2003). The images and forms of these buildings circulated widely and were available to stand in for being modern (and American) in a wide range of different contexts, including within the skyscrapers themselves (see Figure 3.1). New York was consciously, if quickly forgetfully, borrowing

Figure 3.1 Representation of the Empire State Building within the building, including Aztec-inspired sun motif and celebration of New York. (Source: courtesy of Steve Pile.)

from European innovations in design and building form, as well as from much older Renaissance Italian forms such as the campanile.[1] Working with the same 'traditional' skyscraper form of base and roof ornament draped with a curtain wall, moderated by the zoning envelope introduced in 1916, the prolific borrowings of the art-deco designers fashioned an ornamental style for the new skyscrapers from various 'exotic' or 'primitive' styles around the world (Robinson and Bletter 1975). This style spoke so confidently of the modernity of the places doing the borrowing that the many different sources of inspiration for these buildings have failed to register in subsequent formulations of the origins of urban modernity. Benton (2003) sums up the range of influences on New York's architectural ornamentation in this period, when he notes that it 'derived from a wide range of sources, including French Art Deco motifs, Frank Lloyd Wright's Midway Gardens in Chicago, Moorish pattern, pre-Columbian Art, German Expressionism and abstractions from classical and gothic detail' (2003: 246).

In architectural design terms, then, as well as in terms of the hugely diverse origins of the various populations and cultures that make up the city, New York is as much a Mexican, Egyptian, African and Oriental city as it is an icon of the West (see Douglas 1996: 448–9 for more detail; see also the sun motif in Fig. 3.1). But in a rubric of being modern which exemplifies and valorises the apparently autonomous modernity of certain places these circulations and thefts, although at times troubling to the confidence of design professionals such as Mumford, were forgotten as a supposedly 'Western' or 'American' modernity travelled the globe. Thus, even as the major motifs of art-deco design – 'cantilevered "eyebrows" over windows and doors, banded windows, fins, ribs and moulding strips, as well as many kinds of abstract decorative ornament' – 'were endlessly repeated all over the world' (Benton 2003: 248), their diverse origins were defeated by their confident signification as 'Western' modernity, even when in their own cosmopolitan travels they were transformed and adapted to local needs or traditions in different places.

What must be seen, then, as the 'ambiguous'[2] or borrowed modernity of New York is perhaps exemplified by the designs for the 1939 New York World's Fair. The dominant architectural form in the commercial pavilions relied on an art-deco/moderne style drawing on the work of American architects and industrial designers closely associated with the style (Benton 2003: 254). With its loose and easy borrowings from historical and popular sources as well as futuristic industrial design, the 'art deco' idiom was well suited to the first world fair to frame participation in commercial rather than national terms, as large corporations rather than governments dominated the prominent pavilions (Nye 1992: 2). Moderne had become the style of choice for world fairs at this time (Rydell 1993) – to the distaste of architectural critics from Chicago to Johannesburg and certainly to the dismay of those in New York (Santomasso 1980, Benton 2003: 254). For here in the view of the internationalist critics, compared to the popular and commercial designs of the US pavilions, the Brazilian contribution to the fair stole the show as one of very few examples of avant-garde international style modern architecture (Figure 3.2).

Though for Mumford the critical attention given to the international style during the fair undermined confidence in the uniquely 'American' art-deco architecture, it also pointed to the coexistence of a range of different historical and traditional styles within this supposedly quintessentially modern city. Alongside the moderne and modern styles on display, the layout of the fair as a whole exemplified traditional classical and *beaux-arts* planning principles. And across the city more generally, beneath the 'toploftiness' and hype of skyscrapers – in fact on street level – the city of New York carried a diversity of built forms that embraced the monumental and horizontal features of the city just as much as they did the vertical. In New York, a corporate verticality vied with civic horizontalism for prominence in urban design (Taylor 1992). More than this, even as the city was growing skyward it took observers some time to shift their appreciation of the buildings from older emphases on the horizontal streetscape. New visual forms emphasising the vertical and appreciating the aesthetic value of the new urban forms took some decades to emerge (Domosh 1990, 1992). Modern New York drew together cultural practices and built forms that incorporated classical and traditional enthusiasms with strong references to unchanging solidity in some areas of urban design.

Like most big cities, then, New York stood for much more than the (unambiguously) modern. The most ostentatiously modern architectural components of the urban landscape here relied on borrowings from distant places, and recast much older cultural references even as they laid to claim to

Figure 3.2 The Brazil Pavilion at the New York World's Fair, 1939. From the collection of Dr William R. Hanson, published by Paul M. Van Dort in the 1939 New York World Trade Fair's photo collection tour.

the invention of novel urban forms. New York's art-deco experiments were internally dependent on other places and other periods for their inspiration; they also sat alongside a very diverse urban landscape. If New York is to be thought of as a site of modernity, what it means to be modern will need to be reframed. The sources of innovation need to be thought of as dispersed – across places like Renaissance Italy, Aztec Mexico, early twentieth-century Europe and contemporary Japan. The times and sites of urban modernity can be imagined as truly cosmopolitan. More than this, though, the many different styles that made New York a distinctive, even iconic, place can all be appreciated as contributing to the form of urban modernity produced there. Looking back on the ageing landscape of modernity in New York, Ward and Zunz (1992) observe that 'the landscape of modernity combined formal and informal economies, tall and small buildings, the service sector and industry, and the deeds of machine politicians and those of reformers' (Ward and Zunz 1992: 5). The massed verticality of the city might have captured both the commuters' fleeting glance and the camera's emerging aesthetic gaze through the twentieth century. But looking beyond the snapshot of the modern city skyline (Taylor 1992: 32–3), this iconic modern city – horizontal as much as vertical, innovative especially in its borrowing – might serve the world better if it circulated in this more diverse, ambiguous and insecure form.

Misplaced, or place of origin? Rio modern/e

Dislocating the modernity of a city such as New York, supposedly central to contemporary Western modernity, places the history and experience of modernity in a city such as Rio de Janeiro in a new perspective. Once New York's urbanism has been understood as just another cosmopolitan and provincial form of modernity, any anxiety on the part of other cities about lagging behind or mimicking a modernity that is always 'elsewhere' (Johnson 1999) becomes less plausible. Certainly New York's modernity came from elsewhere and, viewed from Rio de Janeiro at various moments, New York might be considered to have been rather backward, in its commercial commitment to the populist moderne style for example, and in its slowness in embracing a more contemporary and critical international architecture. Indeed, there were moments when New York critics looked to Rio de Janeiro as an exemplar of innovation and modernity in urban design. The Brazil Pavilion at the New York World's Fair (Fig. 3.2) sparked a wider interest in design in South America and in 1943 the Museum of Modern Art in New York held an exhibition, 'Brazil Builds', showing the contemporary work of over twenty different Brazilian architects (Fraser 2000: 1). Fraser notes that for the decades from the 1930s to 1960s, 'Latin American architecture was valued for its inventiveness and confidence, and for the way it combined modern and national or regional characteristics' (Fraser 2000: 2).

But in Brazil, local narratives about the ambition to be modern were inextricably bound up with the framing of modernity as coming from 'outside' (see, for example, Kirkpartick 2000: 196), and the modern as consequently being 'out of place' (Johnson 1999). During the turn of the century 'belle époque', for example, Parisian urbanism in the classically inspired *beaux-arts* tradition and urban planning as implemented by Haussman profoundly influenced the reshaping of the urban environment in Rio de Janeiro. In the first decades of the twentieth century, wide Haussman-style avenues, invasive regulation of health and housing in the interests of urban redevelopment and grand classical vistas remade central Rio de Janeiro into the showpiece of Brazilian modernity (Needell 1987). Overcoming disease and apparent disorderliness in the capital city was seen as crucial to ensuring that foreign investors and visitors supported Brazil's ambitions for economic growth and modernisation. In this process, where French culture and British business dominated elite activities, Brazilianness, especially in the form of indigenous and African culture, was devalued (although an important part of even elite urban experiences). African, poor and working-class urbanisms, which had generated popular carnivalesque and at times raucous entertainment for all classes, increasingly met with short shrift from the modernising elite (Meade 1997: 41).

The importation of modern styles from other places (Paris as well as London and Buenos Aires) in the production of elite modernity in the city of Rio de Janeiro was a source of delight and a marker of achievement for the elite. They articulated their ambitions to be seen as modern through efforts to transform the city into an attractive place for visitors and for living, in the latest European styles. But the continuing visibility of African and poor *cariocas* served as a painful reminder of Brazil's difference and, for many local observers, disappointingly highlighted the amazing achievements of the belle-époque developments as mere copies of the original modernity of Europe (Needell 1987). Far from being portrayed with confident acclaims of autonomous modernity, as in New York, the cosmopolitanism of Rio de Janeiro's modernising initiatives was contrasted with their economic 'backwardness', the persistence of tropical diseases and the continuing vitality of their African and indigenous racial heritage. These were cast as markers of their continuing dependence on and distance from the norms of European and North American modernity (Stepan 2001).

Moreover, the poor responded to their removal from central areas to make way for prestigious new developments and to the intrusive state attempts to ensure public health with frequent protests and riots (Meade 1997: 62–3). One of these, in 1904 (the Revolt of the Vaccine), took ten days to bring to a standstill, requiring all the available troops to put it down (Sevcenko 2000: 85–9). This period also saw the rapid expansion of makeshift dwellings – *favelas* – on the steep hills near the centre of Rio de Janeiro where, since the end of slavery, the poor had sought a foothold in the centre of the city (Sevcenko 2000). It was the poor who had to suffer the inconvenience and expense of 'making the capital a healthful and beautiful place' (Meade 1997: 74) and the consequent protests and riots 'proved to the Republican ruling class [. . .] that their ideal of a Europeanized, civilized capital remained elusive' (Meade 1997: 169).

This narrative of cultural ambivalence about the achievements and failings of cosmopolitan urban modernity fuelled the search for a more confident register of economic and cultural independence (Ortiz 2000). A few decades later, the strongly nationalist and modernising regimes of the twentieth century provided the opportunity for Brazilian architects to develop a distinctive modern architectural style. The pavilion at the 1939 New York World's Fair, for example, was 'a very clear statement about Brazil as a country of the future' (Fraser 2000: 184). The proliferation of both moderne and modern styles in Rio de Janeiro highlight the inadequacy of a view of modernity as originating in the West and circulating in borrowed form and laggardly fashion around the 'rest' of the world. For in both these styles, the borrowings of Western urbanists from the work of architects in places such as Brazil and Rio de Janeiro were profound – what was labelled 'Western' was already cosmopolitan.

Brazilian architects contributed, through their appropriations and reconfigurations, to the development of both these styles internationally, and indigenous Brazilian cultures were amongst those drawn on in the European avant-garde's turn to primitivism. As Sevcenko suggestively observes of the post-First World War era, 'paradoxically, when the young Brazilian elite went to Paris on their customary annual visit to breathe "the superior airs of civilsation", they discovered, to their enormous surprise, that the latest trends in Paris were based on the despised culture of the marginalised population of their own country' (2000: 90). Certainly, this would have only been one strand of the prolific borrowings of European primitivism, but no doubt for contemporary travellers these discoveries about the Brazilian roots of elements of European modernism would have been striking. Certainly, as Brazilian art deco came to prominence, these European styles recirculated through Brazil, where similar primitivist borrowings from native South and Central American designs became common. A subtle blending of internationalism and national identity formation – key to the configuration of Brazilian identity in this period – could find easy expression in this form of architecture, which became very popular in 1930s' Rio de Janeiro (Cardoso 2003).

Drawing on both the Parisian sources of art deco and the attraction of the skyscraper developments in North America, art-deco urban designs resonated with long-standing Brazilian investments in novelty, progress and national development. As Cardoso notes:

> The skyscrapers of the Jazz Age were at once modern and non-European; and they spoke of a time in which the Americas would come into their own as the new epicentre of world progress [. . .] Like their architectural counterparts many fine and decorative artists appear to have perceived Art Deco as a particularly suitable idiom for reconciling

the traditional opposition between modernity and national identity, however uneasy that co-existence may eventually have proved.

(2003: 401–2)

In addition to these circulations between Brazil and Europe in relation to the Art Deco style and the fashion for primitivism, Fraser (2000) suggests that Latin America also had considerable impact on Europe in terms of the development of the competing modern architecture of the international style. While Le Corbusier inspired early Rio de Janeiro constructions in this style, he was also himself inspired by the Brazilian landscape. It was the Brazilian waterways that motivated him to introduce curvilinear forms into his architectural vocabulary, for example, and his grand designs were inspired by contexts such as Brazil where there were opportunities for massive redevelopments because of rapidly growing cities and elite and authoritarian government (Fraser 2000). Encouraged by Le Corbusier and building on experiments in early examples of the style, Brazilian architects, through the use of local granite, baroque-influenced tiling or tropical vegetation, shaded the international style into a Brazilian modern (Stepan 2001). Indeed, the tension between Brazilian context and international design was brought out during later bitter attempts by Le Corbusier to claim one of the first international-style buildings in the world as his own work, rather than that of the local Brazilian architects who he worked with. He was brought out to Rio de Janeiro to work with Brazilian architects on the Ministry of Education and Health Building as well as the new University Campus in 1936, both of which became iconic examples of international-style modern architecture (Fraser 2000). These personalised disputes illustrate perhaps that origin and copy are much more enfolded than narratives asserting the dominance of a 'Western' form of modernity allow.

But where New York rose easily to assume its independent and universal modernity despite circulating practices and borrowed designs, the post-colonial placing of Rio de Janeiro persistently undermined the confidence of such assumptions there. Critics have grappled with the idea of a misplaced modernity (Schwartz 1992), or a borrowed modernity out of place (Ortiz, cited in Johnson 1999: 189) with the country's slave past, with its backward or underdeveloped economy, with the poverty that has persisted in the low-wage, high-unemployment post-slave economy. But the examples of moderne/modern architecture and urban design open up for us the possibility of a quite different relationship between apparent origin and copy. At times, as we have seen, New York was looking to Rio de Janeiro for inspiration. However, the attention paid by New Yorkers to architectural innovations in the similarly 'new' societies of Latin America faded from the 1960s onwards, when a language of dependency and the 'Third World' shaded Brazilian architectural adventurism as inappropriate. As Fraser summarises in relation to the later architectural innovations in the design of the new capital city, Brasília:

By the mid 1950s, when Brasília was born, the USA had achieved real political and economic strength, and its attentions were directed elsewhere: to NATO and the USSR. By the later 1960s Brasília was the seat of a vicious military dictatorship ruling over economic crisis – by which time, to paraphrase Serge Guilbaut, the USA had stolen the idea of modern architecture, and had stolen it from Latin America as well as from Europe. Europe, too, had recovered its equilibrium, and between them they could afford to bundle away Latin America into one of those new categories of 'Third World' or 'underdeveloped'. In the process Brasília soon came to be seen not as an outstanding achievement, but as an outrageously ambitious project for a country like Brazil.

(Fraser 2000: 254–5)

Claiming the right to be modern, for cities of all kinds need not diminish awareness of the interdependence of any city on a wide range of other places and contexts, nor should it undermine attention to obvious differences in wealth, infrastructural capacity and economic power amongst cities. But reasserting the dynamic modernity of all cities means dispossessing Western cities of the assumption that it is their experiences which determine the character and future of urban modernity. It also means they cannot simply forget that they too are derivative, imitative and that urban innovation is almost always a result of a cosmopolitan interdependence.

APPROPRIATING THE MODERN: BEYOND MIMICRY

Modernization continues to be commonly understood as a process begun and finished in Europe, from where it has been exported across ever-expanding regions of the non-West. The destiny of those regions has been to mimic, never quite successfully, the history already performed by the West. To become modern, it is still said, or today to become postmodern, is to act like the West.

(Mitchell 2000: 1)

If New York and other powerful Western cities can be dispossessed of their claims to being the sources of an originary version of modernity, it should be possible to read the circulations of modernity through cities and nations around the world as other than purely imitative, as immeasurably more than the copy that the Brazilian elite in Rio de Janeiro lamented. The challenge for a cosmopolitan imagination of urban modernity is to reclaim the experience of modernity's circulations not as mimicry but as appropriation. As the previous section demonstrated, if modernities in a world of circulations and borrowings are a product of interdependence, then the creation of modern urbanisms in cities everywhere – rich and poor, powerful or not – reflect a range of diverse borrowings as well as local inventions and histories. But

Kirkpatrick (2000) notes that our narratives of cultural autonomy can distort this: 'As Latin Americans were well aware, historically global flows are presumed to originate in the West, while other cultural flows are rendered invisible within the world system or are characterised as copies' (Kirkpatrick 2000: 181). A cosmopolitan imagination brings into view the range and multi-directionality of cultural flows. It can also help us to reconsider the cultural meaning of the diverse flows and borrowings that are often gathered together under the sign of the 'copy'.

Local enthusiasms and colonial impositions can both do the work of circulating 'copies' of different modernities, for example. But a very large literature on post-colonial cultures suggests that we can generally predict an active cultural politics of engagement with circulating modernities – even when modernity is externally imposed (for example, Appadurai 1996, Chakrabarty 2000, Comaroff and Comaroff 1997). Rather than generating pure copies, Dilip Goankar proposes that modernity's circulations engender a process of 'creative adaptation'. He suggests that we explore the sites 'where a people "make" themselves modern, as opposed to being "made" modern by alien and impersonal forces, and where they give themselves an identity and a destiny' (2001, p. 18). There is a shortfall here, though, since emphasising the 'creative adaptations' of circulating modernities tends to reprivilege and recentre those, usually assumed to be Western, modernities whose flows around the world have most visibility (Mitchell 2000). 'Alternative' modernities are then assumed to arise elsewhere, leaving intact the originary nature of the 'Western' modern. As we have seen, the modernities that name themselves 'Western' are already appropriations and hybridisations. The landscape of modernity can be imagined as more truly multiple so that the 'west' is no longer a necessary referent in the creation of ways of being modern.

More than this, a narrative of creative adaptation relies on the emergence of 'hybrid' modernisms, and misses examples of pure appropriation, usually styled as 'copying' or mimicry. In this account, the 'west' continues to own the products of its modernity – until they are decisively changed through adaptation and hybridisation elsewhere. But the example of the circulations of architectural form through New York and Rio de Janeiro above suggested that borrowings are often styled as mimicry in Rio de Janeiro, and yet stand as indicators of autonomy in New York. In placing these cities side by side, it quickly becomes clear that the cultural landscape of appropriation is commonly denied authenticity in cities marked with tradition, with poverty and with race.

In response, during the early decades of the twentieth century,[3] Latin Americans engaged with the dilemma of a borrowed modernity in a determinedly assertive fashion, claiming 'their right to assimilate whatever they pleased' (Kirkpatrick 2000: 181). Inspired by European primitivism and their own familiarity with aspects of indigenous Brazilian history, a group of avant-garde modernists articulated their right to assimilate, to appropriate and to reuse anything they chose from anywhere. Revitalising a story about a group of indigenous Brazilians who cooked and ate a shipwrecked coloniser, they proposed a politics of cannibalism (Resende 2000: 202; Oliven 2000: 58). The 'Cannibalist Manifesto' was published in the *Revista de Antropofagia*, year 1, no. 1, in the 374th year of the Devouring of Bishop Sardinha, and proposed a mode of cultural appropriation known as 'anthropophagism' that demanded that Brazilians ingest, creatively digest and use whatever they found admirable and exciting from anywhere. Although disconnected from popular movements and easily drawn into a conservative rural politics, the modernists transformed the aesthetics of elite Brazilian culture and promoted native and popular culture, contributing to the strong nationalist imaginations of the post-1930 'Estado Nôvo' regime (Sevcenko 2000). As Jaguaribe puts it, 'No longer imitators of subservient colonial subjects, Brazilians were now cultural cannibals who devoured the more savoury bits of European culture and cooked them together with African and Indian ingredients into an overwhelming cultural concoction' (1999: 302).

The Argentine writer Borges went further; insisting that supposedly foreign cultures were already an intimate and internal part of Argentinian cultural life (Johnson 1999). Invoking the image of a sphere, 'a sphere whose centre is everywhere and whose circumference is nowhere' ('The Fearful Sphere of Pascal', in *Labyrinths*, 1964), Borges proposes an imagination in which there are no peripheries and centres; instead, all peripheries are already the centre, and vice versa. Without such a vigorous critique of the idea of modernity as mimicry, Yúdice argues, poorer and more peripheral contexts are forever 'condemned to be either Europe's civilisational double, or its civilisational other' (1999: 54). In the anthropophagic model explored here, modernity is permanently available for appropriation and recirculation, without loss of the basis for close identification with circulating cultural forms and with equal rights to possess and to recast their meaning and significance. Different forms of urban modernity, then, can belong to anyone, anywhere. In an anthropophagic view, cosmopolitan modernity circulates in many different directions, from many centres to many peripheries, and can be indigenised to taste.

Appropriating verticality

A direct challenge to the ownership of the icons of urban modernity by cities in the West was launched by Kuala Lumpur, Malaysia through the 1990s. Here the Government signalled Malaysian ambitions to be a part of the global economy by building the tallest building in the world (King 1996, Goh 2002: 52). Aiming to benefit from the rapid economic growth and industrialisation in the wider south-east Asian region, and with the inspiring catch phrase '*Malaysia Boleh* – Malaysia can', Mahathir Mohamad, prime minister and motivator, signalled the country's intention to capture – to ingest, appropriate, rebrand – circulating global modernities as Malaysian (Bunnell 1999). Interestingly, the development of the Petronas Towers, while embodying some Islamic references (such as the basic floor plan design, attributed to Mahathir himself), ties Kuala Lumpur to Latin American modernisms of the early twentieth century. The architect was an Argentine, César Pelli, working in New York (and designer of the Canary Wharf building in London); and the gardens at the base of the towers were designed by the Brazilian modernist landscape designer Roberto Burle Marx (Goh 2002: 60, Bunnell 1999: 18).

As with the earlier Brazilian modernisms, Kuala Lumpur's appropriation and indigenisation of international architectural verticality to their own globalising ambitions is not dependent on the hybridisations implied by the light Islamic references in this case, or the baroque and tropicalist references in the case of Brazilian forms. Even Malaysianisation of the skilled workforce – a formal state policy – was only half-heartedly pursued. Malaysianisation of architecture may have shaped the form of dramatic skyscrapers in Kuala Lumpur during the 1980s but, as Bunnell (2004a) argues, with the Petronas Towers nation-building no longer symbolically realigns the internationalist and modern physical forms of these gigantic buildings – rather it is the global ambitions of a modernising Malaysian elite. The intention was to place Kuala Lumpur on the global map of circulating capital – to capture the attention of whoever is out there with money to invest. These appropriations are of a globalised modernity – transnational in its appeal, imagination and construction (Olds 1995, King 1996, 2004). The Petronas Towers stand for that moment in modernity's circulations when the modern – the skyscraper that had become an international sign of economic and political power – has been blithely appropriated by Malaysians for their own ends. The modern skyscraper, the tallest building in the world was, for a time anyway, Malaysian, part of wider efforts, especially in China, to mobilise this architectural form to support assertions of the rise of 'Asia' as the 'leading edge of modernity' (Bunnell 2004a: 116; see also King 2004).

In the city of Kuala Lumpur itself, though, the towers held different meanings for residents who could hardly escape its visual presence. They have been tied into the rumour-filled vacuum of authoritarian politics and the corrupt practices of companies dependent upon state patronage (Bunnell 1999, 2004a, Sardar 2000). But they have also been seen as an embodiment of elite assumptions about appropriate urban ways of life for Malaysians. Poor and traditional Malaysians were identified as not fit for a global city, through the extreme contrast which the towers presented to more informal elements of the Kuala Lumpur cityscape. For some observers, these towers ironically highlight even more strongly the pockets of poverty and residential decay that remain in this rapidly modernising city (Bunnell 2004b). As we have seen in the case of Rio de Janeiro, modernity is often felt to be undermined by the coexistence of tradition, poverty or indigenous practices: in this case, those with a rural *kampung* (village) lifestyle, deemed too traditional for modern urban living (Goh 2002).

However, Malaysia's appropriative modernities of globalising economic development have also found ways to imbricate indigenous tradition with global ambition in the cultural practices of urban modernity. Political discourses about ways of being modern in Kuala Lumpur draw prolifically on a (changing) imagination of a traditional Malay, or *kampung*, way of life. On the one hand, observers might be forgiven for imagining that traditional ways of life are excluded from the official version of the modern Malaysian city, as symbolised by the Petronas Towers. Globalising development initiatives such as high-rise housing estates and rapidly expanding formal employment opportunities leave no place, for example, for social behaviours that seem out of synch with initiatives to create a 'new Malaysian' subjectivity suitable for the country's ambitions for global economic success (Bunnell 2002: 1690). But on the other hand, *kampung* values – of mutual support, community and egalitarian social interaction – are strongly promoted as codes for articulating modernity in distinctly moral and Malay terms and have been recast as symbols of progress even as actual *kampungs*, or informal settlements within the city, are studiously eradicated as 'sites which both signify and propagate inappropriate Malay conduct' (Bunnell 2002: 1690; also Yeoh 2001, Bunnell 2002: 1695; Goh 2002: 54–6).

Sardar notes that 'the post-independence history of Kuala Lumpur has been the translation, in a particular form, of the *kampung* into the city and the nation' (2000, p. 158). Traditional practices have fed into government visions of contemporary Malaysian modernities and, more consciously so, in recent years as 'the image of the Malay village, or *kampung*, has changed from a symbol of Malay backwardness to an idealized repository of pristine Malay culture and values' (Goh 2002: 49). This dynamic has shaped the statist 'managed democracy' that operates through patronage networks and is influenced by a search for traditional status characteristic of both the communal organisation of Malay society historically and the political culture of the ruling party (Jeshudason 1995, Sardar 2000: 160). But equally, outside of formal political ambitions, traditional practices continue to shape life in the city, in the informal, rumour-mongering critique of powerful actors and the resilience of local cultures – whether in the informal food markets that spring up around new high-rise developments, for example, or *kampung*-style settlements and social relations (Sardar 2000, Thompson 2002: 70). These alternative ways of being modern and urban challenge both political authoritarianism and the relentlessly ambitious forms of urban redevelopment, which have remade the landscape of Kuala Lumpur (Goh, 2002). As Evers and Korff note:

> The modernity of the Southeast Asian city is thus not economic growth, industrialisation and the emergence of modern high-rise quarters and housing estates, and underdevelopment is not subsistence production, the slums and *kampung*. The modernity consists of both the slum and the high-rise, subsistence production and global finances.
>
> (2000: 23)

If scholars are to theorise urban modernity in more cosmopolitan ways, this diversity of urban experience needs to inform analyses. Whereas tradition, poverty and racial difference have historically marked certain cities and people as not-modern, in a post-colonial urban studies modernity belongs to all places and people. The internationally circulating high-rise building has been energetically appropriated to symbolise economic success in cities around the world (King 2004); but the dynamic *kampung* culture of Kuala Lumpur reminds us that in many cities modernity is at least doubled. This was the case in New York, where the internationally famous skyline of the 1930s was matched by the prolific international circulations of the work of African-American artists, novelists and musicians (Gilroy 1993). And in Rio de Janeiro, the modernities of popular music and performance – associated more with African and indigenous cultures – competed with the success of the internationalist architects in shaping the city's international reputation. Modernity, in the shadow of colonialism and slavery, is often experienced in 'black and white', fracturing along racialised divisions. This offers an important explanation for the exclusions of many cities around the world from scholarly accounts of urban modernity. But the reach of colonialism in contemporary urban studies, as we have seen, has been refracted through the divisions of developmentalist thinking. The following section turns to reclaim experiences of urban modernity – and the invention of modernities – for cities that have for too long been excluded from such accounts by virtue of their poverty.

MODERNITIES OF UNDERDEVELOPMENT?

> 'At sight of new styles you always discard your old customs and nurse the new' (traditional Tswana proverb, from South Africa[4])

Inventing ways of being modern and urban in Rio de Janeiro and Kuala Lumpur takes place in contexts of division and inequality. As we have seen, this has tended to mark these inventions, setting them apart from those of wealthier and Western contexts. Not that poverty and inequality are uncharacteristic of the iconic locations of American urban modernity; for example, Chicago and New York, in their different ways, divided and segmented both wealth and cultural invention throughout the twentieth century (Abu-Lughod 1999). But in an urban studies after colonialism and developmentalism, a range of hierarchising imaginaries and practices have ranked some cities behind, or below, others in terms of levels of cultural and economic development. It has also been imagined that these different cities foster 'backward' ways of being urban, or modern. Marshall Berman (1983), for example, explores what he calls 'modernities of underdevelopment' in the context of St Petersburg before the Russian Revolution, where he observes that:

The anguish of backwardness and underdevelopment played a central role in Russian politics and culture, from the 1820s well into the Soviet period. In that hundred years or so, Russia wrestled with all the issues that African, Asian and Latin American peoples and nations would confront at a later date. Thus we can see nineteenth-century Russia as an archetype of the emerging twentieth-century Third World.

(Berman 1983: 175)

The modernisms of underdevelopment, Berman suggests, produce strange versions of urban modernity. He describes these as variously 'warped', 'weird', 'unreal', 'surreal', 'illusory', 'unbalanced', 'bizarre' or 'dream-like' urban forms and cultures (1983: 181, 193, 205, 249, 272, 284) – making St Petersburg into a 'mirage, a ghost town, whose grandeur and magnificence are continually melting into its murky air' (1983: 192). Such views of the diverse experiences of modernity label them as different, even deficient. For Berman, the different modernity of St Petersburg was marked by its poverty and peasant economy, with a thin veneer of modernity barely hiding other ways of life.

The city of St Petersburg was heroically built from scratch on marshland in the early decades of the eighteenth century following the very best of contemporary European design – only on a larger scale and rather more extravagantly. But, like Rio de Janeiro at the turn of the twentieth century, St Petersburg had to contend with the contrasts between this grand urban environment and poverty, the terrible living conditions of working people in the city, an economy largely dependent on peasant agriculture, as well as the persistence of status hierarchies and aristocratic traditions within the new urban society. All these undermined the 'illusion' of modernity and commercial democracy generated on the 'Nevsky Prospect' – a shopping street and boulevard at the centre of the modernising dreams of St Petersburg's founders and stage for 'the encounter between Russia and the West' (Berman 1983: 194, fn 12): 'Thus the Nevsky is a kind of stage set, dazzling the population with glittering wares, nearly all imported from the West, but concealing a dangerous lack of depth behind the brilliant facade' (Berman 1983: 230).

For Berman, then, 'in relatively backward countries, where the process of modernization has not yet come into its own, modernism, where it develops, takes on a fantastic character, because it is forced to nourish itself not on social reality but on fantasies, mirages, dreams' (Berman 1983: 235–6). This infuses cultural and political practices in the city 'with a desperate incandescence that Western modernism, so much more at home in its world, can rarely hope to match' (Berman 1983: 232). To Berman, the St Petersburg situation was somehow more crazy because, as some observers in Rio de Janeiro had noted, the fantasy of Western modernity seemed to offer only a thin veneer over the 'backwardness' of actual urban experience (Berman 1983: 229–231).

Berman's hope is that his 'trip through the mysteries of St. Petersburg, through its clash and interplay of experiments in modernization from above and below, may provide clues to some of the mysteries of political and spiritual life in the cities of the Third World – in Lagos, Brasilia, New Delhi, Mexico City – today' (Berman 1983: 286). Although for much of his analysis Berman is suggesting that the modernities of these places are distinctive, unusual, even weird, he does eventually place the history of modernism in Petersburg within the same time-stream as cities of Western and wealthier contexts. He hopes that there is something to learn from Russia for all modernisms, which is to see Petersburg 'as the archetypal "unreal city" of the modern world' (Berman 1983: 176). Despite his characterisation of modernities of underdevelopment as quite different from those in wealthier places, he acknowledges that they do speak to other contexts, and even contribute forms of modernity that circulate and travel elsewhere.

Specifically, Berman suggests that St Petersburg was the site for the invention of one of the primary genres of literary modernity – a literature of and about life on the city street. In the brutally repressive political context of tsarist Russia, and within the stifling constraints of aristocratic social conventions, the street offered crucial opportunities for contesting power relations even if through fleeting, demonstrative encounters or events. Berman observes the birth of a kind of modernism from below – 'a movement of Petersburg plebeians striving, in increasingly active and radical ways, to make Peter's city their own' (Berman 1983: 235). Through the history and literature of St Petersburg, Berman suggests, the inventive writers and ordinary citizens offer examples of ways to understand and engage with the 'unreal reality of the modern city', to confront and challenge forms of power, to become 'both personally and politically "more alive" in the elusively shifting light and shadow of the city streets. It is this prospect above all that Petersburg has opened up in modern life' (Berman 1983: 286). He feels that this insight will be useful and relevant in all contexts where there is a 'clash and fusion' of different modernities – and he suggests this is true of cities from New York to Tokyo, Madrid to Mexico City.

All cities, then, are sites of clashing and contestation, and stages for the ephemeral reconfiguration of meaning on the streets. Rather than being a feature of underdevelopment, cities are perhaps intrinsically unreal, weird and illusory as a result of the diversity of things and people they gather together. Finding ways of being modern, of being urban, is often done in a context of stark contrasts, or in a mode that is out of step with some other aspects of the city. Following Berman, then, there is much that could be learnt by those in wealthier contexts from understanding ways of being modern in poor places. But this can only be achieved if we abandon the temporal dislocations that place poor societies in a different time zone from wealthy ones. 'Backward', 'developing', 'primitive' and 'traditional', even 'underdeveloped', are terms that redistribute historical time across geographical space – and that prevent understandings of urban cultural inventions in different contexts from informing one another. To avoid this source of incommensurability amongst accounts of modernity in different cities, all cities need to be understood as coeval, as existing in the same time (Fabian 1983).

The transitory politics of life on the street, ephemeral, sporadic attempts at contesting brutal power and fleeting attempts to appropriate and reuse the city to make new kinds of political and cultural meanings all feature in the modernisms of St Petersburg, in what Berman calls the modernities of 'underdevelopment'. And it is on the street, after all, that in poorer contexts much of contemporary urbanisms are enacted. In situations of limited infrastructural development, poor housing and insecure livelihoods, the street, and other 'ephemeral' public spaces (Simone 2001) become crucial settings for inventing ways of being sociable, ways of securing opportunities to earn money and ways of gaining recognition. The inventiveness and modernity of city life everywhere is not only in bricks and mortar, but also in personal styles, performances and cultural practices. Between botanising the asphalt with Baudelaire in Paris, bumping shoulders on the Nevsky Prospect in Petersburg, and looking decent on Cairo Road in Lusaka (which we will turn to now), there may be less distance than might be imagined.

Disappearing the West

Reusing and living with old modern structures or faded and decaying infrastructures is by no means a distinctive feature of poor cities, as we noted in respect of New York (Ward and Zunz 1992). Built in to modernity's ambitions to be novel and contemporary is the inevitability of becoming outdated. Jaguaribe (1999) considers the ruin of Rio de Janeiro's first modernist skyscraper in the context of what she calls the 'unconfined city'. Tropical climate and poor maintenance have left the famous Ministry of Education and Health Building in a state of some decline (Stepan 2001), but the modernist ruin has also been reabsorbed into the diverse imaginations and practices of the multicultural city. As she comments:

> In the contradictory pulses of the city, the modernist ruin dismantles its utopian project and becomes a monument that re-symbolizes our historical trajectories by fabricating a myriad of dialogues with the recent past. The aged functionalised building projects its dismantling silhouette against the changing city. Modernist tiles contrast to the kitsch trinkets sold by street vendors. The unconfined city transforms architectural forms and national narrations into a plurality of histories.
>
> (Jaguaribe 1999: 312)

Previous elite projects to regenerate the city, to signify progress and nationalist achievements have become reconnected to the liveliness and livelihoods of the wider city. In Johannesburg, South Africa, high-rise modern flats built for white elites in the 1950s and 1960s have been recycled for use initially by previously excluded black South Africans and more recently by foreign Africans (Morris 1999). Residences are managed and inhabited in new ways, sub-letting has become an important component of immigrant economic circuits, and public spaces – street corners and parks – have been appropriated as settings for networking, socialising, eating, trading and simply signposting a presence in Johannesburg (Simone 2004). Street corners, unused public spaces and abandoned structures have become crucial – and contested – resources in inventing livelihoods and securing goods on a small income.

In Lusaka, Zambia, reusing and restyling clothing has become a site for the production of distinctive cultural practices of urban modernity in a time of serious economic decline. Lusaka was initially built as a modern 'garden city' in the 1930s to administer Northern Rhodesia. It expanded during the boom years post-independence, with opportunities for government employment and some state-supported industrial activities (Hansen 1997, Myers 2003). Since the early 1970s, when the price of copper, Zambia's main export product, declined, and especially since the structural adjustment policies implemented by the democratic post-Kaunda government in the early 1990s, formal employment has been substantially reduced. Incomes have been vastly devalued in relation to the cost of formally produced goods. The price of basic commodities has skyrocketed with the ending of government subsidies and nationalised production, and many households have turned to informal trading to attempt to make a living (Young 1988, Moser and Holland 1997).

In this situation, getting by has demanded a reconfiguration of aspirations and of practical strategies for living in the city. What Ferguson (1999) has called Zambians' 'expectations of modernity' have had to be adjusted. No longer associated with industrialisation, formal employment or much in the way of bricks-and-mortar constructions, the creativity and contemporaneity of city life are to be found very much on the streets. Under colonial rule, the performance of individual identity in public spaces had been an important way of countering the low status that white society accorded to Africans in the region (Epstein 1964, Mitchell 1968). Other forms of monetary investment in improving personal well-being and cultural standing were forbidden – owning houses or land, for example. For African people who worked in the city, then, clothing and self-presentation were the most important sites for proclaiming and actively working out ways of living in the city, and being modern.[5] With economic crisis, this is again true and, as Karen Hansen observes, 'we need to reckon with people's preoccupations with clothing if we are to understand the process of becoming modern in this part of Africa' (2000: 15).

The formal sources of clothing that workers enjoyed making their own both under colonialism and after independence are now beyond the means of most Zambians. Instead, many Zambian consumers bring together informal tailor-made clothing in an 'African' idiom (with textiles and styles borrowed from places such as Nigeria and Ghana) and a style known locally as *salaula* (a Bemba term meaning to pick, or to rummage) refashioning second-hand clothing from wealthier countries. A part of Zambian life in pre-independence times, especially through centuries-old trading routes with Zaire and other places in the region, the trade in second-hand clothing has become a major activity in post-structural adjustment times.

But even as Zambians remodel and appropriate the discarded clothing of wealthier countries, they do so in a style that is all their own. As Hansen notes, 'Save for the origin of *salaula* garments, there is nothing particularly "Western" about how people in Zambia deal with them' (Hansen 2000: 251). 'The popularity of imported secondhand clothing in a country like Zambia provides evidence of much more than dressing bodies in a faded and worn imitation of the West' (2000: 5). In Zambia, for example, the clothes are determined 'new' if the bales in which they arrive are unopened, they are imaginatively linked to the 'outside' largely through the extensive donor relations that Zambia has with many other countries, and they are worn in combinations that bear testimony to the wide range of external influences on contemporary Zambian style, including Zairean, west and east African, colonial, South African and British heritages as well as contemporary fads in the USA that arrive in Zambia through television, music and other media: 'There is a multiplicity of heritages at work here with complex dialectics between local and foreign influences, and between what is considered to be "the latest" and what is current, in a reconfiguration process that generates distinct local clothing-consumption practices' (Hansen 2000: 253).

A lot of work goes into the reuse of clothing towards producing a Zambian way of being modern – and attention to Zambian norms is crucial, whether these be expectations regarding the presentation and decorous covering of bodies, or a search for uniqueness so as to stand out in a crowd, or a desire to be properly turned out, as office workers or teachers perhaps might be expected to be. Locally generated trends and fashions – 'the latest' – circulate in a performative and sociable context on streets, in the market, or at social events, shaping people's *salaula* selections and clothing compilations. An important link is made for many people, Karen Hansen explains, between understandings of well-being and clothing, with the alternative of 'wearing rags' working as a metaphor for 'lack of access' (2000: 15) to the benefits of development. The styling of bodily appearances on the street, then, is as much about needs as it is about wants; as much about development as it is about fashion.

In the recycling of the objects of modernity, things such as clothing function not as referents to a distant, superior, more modern way of being. In the Zambian case, the 'Western' character of the used clothing all but disappears in the myriad of local meanings and improvisations that produce a distinctly Zambian urbanism. Rather, through their appropriation, circulating objects and fading infrastructures are 'made new', they are created as a part of dynamic modernities, often produced in conditions of crisis and poverty. If the 'west' can be disappeared as a source of modernity in the course of recasting circulating objects in local idioms, then we will need to turn our attention to ways of being modern in cities around the world that draw little or no inspiration from dominant 'Western' circulations of ideas, attitudes or objects. To do this, we will need to definitively abandon accounts of modernity and cities that pitch the achievements of urbanism against the limitations and backwardness of countryside and tradition.

Modern traditions/traditional modernities

Inventing ways of being modern in cities might be partly a response to the conditions of city living – the hardships of life in *favelas* and townships, shaped by economic exploitation and social exclusion have, at times, provided intensely creative environments for music, dance and fashion, for example. But they have just as often depended on the city's ability to collect together diverse circulations and influences. These influences have included bits and pieces of what have been called 'Western' circulating modernities or dominant forms of globalisation, but they also included elements of cultural practices that travel back and forth from the countryside or from neighbouring countries – practices that maintain strong continuities with durable and dynamic indigenous cultures. The sources of inspiration for making oneself 'at home' in different cities, for exploring alternative ways of city life in difficult circumstances, are usually multiple. As Comaroff and Comaroff (1997: 34) note, 'when looked at up close, the modalities of the modern turn out to be multiple, fragmentary, and cross-bred.'

Even the motivations for invention and experimentation, so strongly associated with modernity, are not simply the product of the circulating (Western) modernities of colonialism or globalisation. The Tswana proverb at the beginning of this section reminds us that an enthusiasm for newness permeated many different societies. As migrants relocated to urban areas, they often brought along with them an ability to create new ways of being in contexts of rapid change. Not only did migrants draw on the cultures of their home places to make sense of life in cities – 'activat(ing) a new world of experience in which the gap between country and town is bridged' (James 1999: 37). They also drew on long-standing habits of incorporating new and diverse experiences and artefacts into existing cultural repertoires.

In the northernmost regions of South Africa, Sotho speakers developed a 'traditional' genre of music, singing and dancing, called *kiba* (meaning 'to beat time and stamp') which has become, at different times and in quite divergent ways, important in the cultural activities of men and women migrants from this region who live on the mines and in Johannesburg (James 1999). While migrant men have for many years used this style of performance to consolidate local and village-level home-based groups in workplaces in the city, women have drawn on it more recently (in the past twenty years) to find companionship and support in an urban context. Unlike men, most African women in Johannesburg work in isolated situations and amongst people from many different backgrounds and languages. Their *kiba* groups thus actively reconstruct identities to produce a sense of belonging across a much wider regional and language area. For women, the dance groups represent 'a relationship constructed less out of a shared rural background than out of shared urban experiences' (James 1999: 48): 'Women's *kiba*, then, while apparently harking back to a shared *setšo* [tradition], spoke more of a culture women had acquired from their friends in town, or at least imbued with new significance there, than of any culture they had personally transported intact from the countryside' (James 1999: 62).

The scattered origins of women migrants means that they have to work hard to provide 'a unifying sense of a rural home' (James 1999: 61). Women also use the 'traditional' genre to express how they have had to reconfigure existing expectations of women – assuming more of the position of 'brother' in the network of family obligations, rather than that of dependent wife or daughter-in-law. As their families faced growing economic stress and were stretched across urban and rural locations, many women have assumed more responsibility for their natal home rather than become part of a new marital homestead. Women especially have therefore 'used apparently traditional music as a means of expressing their rather less than traditional outlook on life' – using *sesotho*, or tradition, in new ways (James 1999: 45)

As Deborah James explains, despite 'remarkable continuities of musical form', the *kiba* genre is itself 'not a static or unchanging one' (1999: 71). Its basic structure makes it a (traditional) site of constant innovation: new events and commentaries are constantly incorporated within familiar idioms and previous performances are drawn on in new ways to provide durable links with the past. Performers stress that *kiba* was imported to the Sotho-speaking areas from neighbouring polities or from 'the North' (James 1999: 73) and that as a peripheral form of 'traditional' performance within Sotho culture, it was therefore free to be changed and adapted, for example in migrants' activities, because it was beyond the purview of rulers. New styles also came from other areas from time to time and were imported into villages where they were thought of as 'fashions': 'They swept through the countryside, and often contained lyrics of great national topicality alongside those referring to more domestic and area-specific concerns' (James 1999: 73). Thus while *kiba* is seen in both rural and urban areas as a signifier of 'tradition', intrinsic to its form is the constant reconfiguration of words, clothing and elements of the dance performance. According to James, this lively inventiveness has 'not intrinsically altered musical content' (1999: 75), even as older songs are being revitalised with new material. Indeed *kiba* itself is now circulating beyond both its regional context and the city of Johannesburg, with performances on national television and visits by performers to international music festivals as part of a wider 'world music' movement (James 1999: 6).

The inventive, skilful and deft ways in which many 'traditional' cultures have managed the 'encounter' with different, often dominating and hostile travelling cultures for centuries surely demands that the fantasy be permanently laid to rest that innovation and a fascination with the new belong to Western modernity, in contrast with stable, unchanging traditional cultures. Rituals and traditions, then, are better understood as transformative, creative and constitutive forces in modern life (Comaroff and Comaroff 1997: xvi). Twentieth-century Western urbanism's convenient fiction appropriating creativity for certain 'modern' cities by associating a static version of tradition with poor cities and their countrysides is just that, a fiction. A post-colonial urban studies would frame the city as a site for the constant reinvention of (already inventive) traditional practices such as *kiba*. It would also affirm the coexistence of more or less durable cultural forms in the contemporary moment: 'the latest' and more long-standing practices would both be understood as constituting features of modern urban life. In fact, as we saw in Benjamin's analysis of Parisian urbanism, modernities are often forged in the familiar idioms of older, established ways of doing things (see also Ogborn 1998). Inspired by Benjamin's dialectic of modernity and tradition, then (as we explored in Chapter 1), we can expect to find both modern traditions, and traditional modernities in cities around the world.

The experiences of poor people, or cities in poor contexts, do not stand for the 'past' of Western urbanism – either as sites of tradition or economic backwardness – but are contemporaneous social formations that have produced diverse and distinctive ways of being urban. With inventive traditions and appropriated copies, urbanisms in poor places stand alongside other ways of being modern and can illuminate experiences of city life in other contexts. But, as we will consider briefly in the conclusion, if poor cities are not the 'past' or simply 'copies' of Western urbanism, neither, I suggest, are they its future.

CONCLUSION: NEITHER ARE WE YOUR FUTURE

This chapter has operationalised a cosmopolitan tactic for postcolonialising urban theory. In this cosmopolitan approach to urban modernity, all cities can be understood as both assembling and inventing diverse ways of being modern. Forms of urban modernity everywhere are as likely to be borrowed as created anew; as likely to absorb or to adapt durable cultural forms as to abandon them for new ones. Moreover, as we saw, the circulation of the various artefacts and practices of modernities around the globe inspires their prolific appropriation and incorporation into any number of different ways of being urban.

There remains one conceptual manoeuvre in thinking about urban modernities that we have yet to consider, though, and that is a recent tendency to see the urbanisms of poor places as portraying the future of wealthier cities. In a discussion of neo-liberalism and cities, Neil Smith suggested that 'the leading edge in the combined restructuring of urban scale and functions [. . .] lies in the large and rapidly exploding metropolises of Asia,

Latin America, and parts of Africa [. . .] These metropolitan economies are becoming the production hearths of a new globalism [and] [. . .] progenitors of new urban form, process, and identity' (2002: 436). In an effort to shift the attention of global city theorists from the sites of finance capital to the role of a range of cities in the 'global production of surplus value' (2002: 446), Smith also sets these cities up to figure the future of Western cities. They become, in a sense, figured as hyper-modern. Never having had effective welfare and social reproduction facilities and, often, more recently sites of rampant forms of neo-liberalism that devastated what collective reproductive forms existed, these places are thought to foreshadow the forms of social reproduction likely to be characteristic of an emerging new global urbanism. Both production hearths and prefigurations of coming forms of urbanism, these cities come to represent the future of cities everywhere: and for Neil Smith this future doesn't look good (except that perhaps it will stimulate more contest and struggle over the terms of urban reproduction). The different urban experiences embodied in managing apparently sprawling metropolises of poorer countries within tight financial constraints, structurally adjusting national economies and stretched administrative resources are drawn on to represent the inevitable end point of neo-liberal urbanisms in the wealthier cities of the world.

In this argument, the time chart of urban theory flips neatly around, from setting urban experiences in places such as Africa and South America into the West's past (as traditional, tribal, primitive) cities in many poorer places now commonly configure anxieties about the fate of urbanisation and urban living. Ironically, then, the 'lack' of development is transfigured from a marker of backwardness into a vital indicator of the futures that cities everywhere might face. Depleted infrastructure and 'desperate resilience' (Smith 2002: 436) in the face of economic crisis seems to capture the shape of this noir modernity. As Koolhaas et al. (2000: 653) observe in their reflections on Lagos:

> We are resisting the notion that Lagos represents an African city en route to becoming modern. Or, in a more politically correct idiom, that it is becoming modern in a valid, 'African' way. Rather, we think it possible to argue that Lagos represents a developed, extreme, paradigmatic case study of a city at the forefront of globalizing modernity. This is to say that Lagos is not catching up with us. Rather, we may be catching up with Lagos.

Inverting the problematic of spatialised temporalities associated with ethnocentric views of modernity does little to place diverse cities in relations of temporal equivalence. Instead it continues to rest on the supposition that poor cities do not seem to have achieved the features considered 'urban' in the West. The 'us' of the Koolhaas collective begrudgingly assign urbanity to Lagos, 'for want of a better word' (2000: 652). At the moment when

Lagos, one of the largest cities in the world – estimated to be the twentieth largest urban agglomeration in 2003 (United Nations, 2004) – is slipping off the conceptual register of urbanity, we can be sure that the task of reconceptualising urbanism is much overdue.

The following chapter builds on this need to appreciate a diversity of ways of being urban and the potential benefits of a more cosmopolitan basis for urban theory, specifically in relation to accounts of the impact of globalisation on cities. Theories of global and world cities offer an important opportunity to overcome some of the divisive theoretical manoeuvres that have kept the experiences of different cities apart within the field of urban studies for so long. They take the strong networks that have developed amongst cities as their primary focus. There is a renewed interest in understanding processes that are affecting cities in quite divergent economic and cultural contexts. But even as these approaches have brought into view new transnational economic processes, stretching across a range of different cities, the chapter will consider how they have, perhaps unwittingly, reinstated damaging hierarchical assumptions about relationships amongst cities. In doing so they have also contributed to the popularity of a policy narrative which suggests that even the poorest cities need to face the challenge of building capacities to globalise their economies in order to become more like successful 'global' cities, apparently at the top of a worldwide hierarchy of urban centres.

Just at the moment when a post-colonial critique might allow us to build an account of diverse urbanisms and ordinary cities, and despite their initial promise in this regard, analyses of 'global' and 'world' cities have reinscribed hierarchies and categorisations into the field of urban studies. They have also contributed to the generation of prescriptions for all cities based on the experiences of very few. So in spite of the strong transnationalism of global- and world-cities approaches, the challenges for a post-colonial urban studies remain: to refuse an account of cities that rests on the experiences of only a few privileged centres and to find ways to appreciate the distinctiveness of all cities.

4 World cities, or a world of ordinary cities?

INTRODUCTION

Globalisation has transformed urban studies. Cities are now routinely viewed as sites for much wider social and economic processes, and the focus for understanding urban processes has shifted to emphasise flows and networks that pass through cities rather than the territory of the city itself (Friedman 1995a, Sassen 1991, Smith 2001). The study of cities now commonly encompasses the flows of global finance capital, the footloose wanderings of transnational manufacturing firms, and the diverse mobilities of the world's elite alongside diasporic and migrant communities from poorer countries. In many ways, urban studies has become much more cosmopolitan in its outlook. Globalising features common to many cities around the world encourage more writers to consider cities from different regions, as well as wealthier and poorer cities, within the same field of analysis (Marcuse and van Kempen 2000, Scott 2001, Graham and Marvin 2001). This is certainly good news for a post-colonial urban studies, eager to bring different kinds of cities together in thinking about contemporary urban experiences.

The situation appears more propitious than ever, then, for an integration of urban studies across long-standing divisions of scholarship, especially between Western and other cities, including 'Third World' and former socialist cities. An analytical focus on the transnational global economy could ensure that such categorisations of cities will no longer be of any relevance. Indeed, this is a claim made by the key advocates of these approaches (Sassen 1994, Taylor 2001). Does this mean that urban studies has come to be sensitive to the diversity of urban experiences, to the wide range of cities across the world? Could this be the basis for a post-colonial urban theory that refuses to privilege the experiences of some cities over those of others?

Many studies of globalisation and cities have drawn on the idea of 'world' cities to understand the role of cities in the wider networks and circulations associated with globalisation. Some cities outside the usual purview of Western urban theory – 'Third World cities' – have been incorporated into these studies in so far as they are involved in those globalising processes considered relevant to the definition of world cities. This is definitely a positive development in terms of ambitions to post-colonialise urban studies, to overcome the entrenched divisions between studies of 'Western' and 'Third World' cities. But many cities around the world remain 'off the map' of this version of urban theory (Robinson 2002a). And despite the relative inclusiveness of the focus on globalisation processes, developmentalism continues to pervade global- and world-city narratives, consigning poorer cities to a different theoretical world dominated by the concerns of development. Although the older categories of First and Third World may have less purchase, world cities approaches have a strong interest in hierarchies, and have invented new kinds of categories to divide up the world of cities. Perhaps most worrying for a post-colonial urban studies, world-cities approaches, by placing cities in hierarchical relation to one another, implicitly establish some cities as exemplars and others as imitators. In policy-related versions of these accounts cities either off the world-cities map or low down the supposed hierarchy have an implicit injunction to become more like those at the top of the hierarchy of cities: they need to climb up the hierarchy to get a piece of the (global) action. Being one of the top-rank global cities can be equally burdensome, though, encouraging a policy emphasis on only small, successful and globalising segments of the economy and neglecting the diversity of urban life and urban economies in these places.

So while there is much to learn from global- and world-cities approaches, this chapter suggests that there is still considerable work to be done to produce a post-colonial form of urban studies relevant to a world of cities, rather than simply for selected 'world cities'. Noting especially the adverse political consequences of analyses that emphasise hierarchies and categories and that still divide the field of urban studies along developmentalist lines, the chapter presses the importance of letting all cities be ordinary. World cities approaches, it will be suggested, operate to limit imaginations of possible urban futures, especially in relation to poorer cities, and the situation of poor and marginalised people in cities around the world. A post-colonial urban studies needs to move beyond categories and hierarchies and to abandon claims to represent some cities as exemplars for others. It needs to be able to be attentive to the diverse experiences of a world of cities. While global- and world-cities approaches have much to offer, ultimately they leave these challenges unmet. Instead, this chapter makes the case that all cities should be viewed as ordinary, both distinctive and part of an interconnected world of cities. The last section spells out what this might mean. And, as we will see, an ordinary-city approach is as important for the wealthiest ('global') cities, as for the poorest.

GLOBAL AND WORLD CITIES

World cities are thought to be different from others to the extent that they play an important role in articulating regional, national and international economies into a global economy. With the rise since the 1970s of trans-national investment flows and the emergence of a new international division of labour based on the restructuring of manufacturing production processes, now involving integrated production processes dispersed across the globe, scholars have been drawn to rethink the role of cities in relation to the global economy. It was suggested that some cities were increasingly serving as the organising nodes of a global economic system, rather than simply being linked into local hinterlands or part of national systems of cities. At the same time, many populations were being excluded from these new spaces of global capitalism, and thus from the field of world cities: to these writers some cities were becoming 'economically irrelevant' (Knox 1995: 41).

A corollary, and drawing heavily on wider accounts of uneven development of the world economy, was the suggestion that these global nodes, or world cities, could be arranged hierarchically, roughly in accord with the economic power they command. A key mechanism in making this a dynamic hierarchy, rather than simply a static given was the existence of competition between world cities, which would drive some to rise, others to fall. Furthermore, external shocks and wider structural processes would shape the respective fortunes of world cities and, to some extent, determine their position in the hierarchy. It was assumed, then, that cities could rise and fall through the hierarchy, and that their overall position in relation to other cities would be determined by the relative balance of global, national and regional influence that they could mobilise (see for example, Hall 1981, 1966, Knox and Taylor 1995).

World-systems theory, more generally, which was a strong influence on some early world-cities theorists, proposed that countries across the world are seen to occupy a place within the hierarchy of the world economy and possibly make their way up through the categories (core, periphery, semi-periphery) embedded in the world-economy approach. Following this, the world-cities approach assumes that cities occupy similar positions and have a similar capacity to progress up or fall down the ranks. The country categorisations of core, periphery and semi-periphery in world-systems theory were transferred to these initial analyses of world cities, so that the hierarchies of world cities were determined by relative levels of economic development. This meant that writers also found it useful to continue to use development-related categorisations to describe cities' positions in the hierarchy, including the division between First and Third World cities. A common suggestion though was that with globalisation First World cities were assuming some of the characteristics of Third World cities in terms of a growing number of low-wage, service-sector jobs and poor living conditions – and that some Third World cities were becoming more like First World cities in terms of the balance of economic activities and the presence of a global elite (Friedman 1995a, Jones 1998). Knox writes, for example, that, 'just as we can see the world cityness of regional metropoli, so we can see the Third-World-ness of world cities' (1995: 15). These categorisations were being used more flexibly then, but remained a very important lens with which to view the world of cities. So, at the same time that world-cities analysts were extending their focus to include poorer cities, they were reinscribing developmentalist assumptions about the hierarchical relations amongst cities.

In a prominent contribution to the world-cities literature, Saskia Sassen (1991, 1994) coined the term 'global cities' to capture what she suggests is a distinctive feature of the current (1980s on) phase of the world economy: the global organisation and increasingly transnational structure of elements of the global economy. Her key take-home point is that the spatially dispersed global economy requires locally based and integrated organisation and this, she suggests, takes place in global cities. Although many transnational companies no longer keep their headquarters in central areas of these major cities, the specialised firms that they rely on to produce the capabilities and innovations necessary for command and control of their global operations have remained or chosen to establish themselves there, including advanced business and producer services, legal and financial services. More-over, it is no longer the large transnational corporations that are the centre of these functions, but small parts of a few major cities that play host to and enable the effective functioning through proximity of a growing number of these new producer and business-services firms (Sassen 2001a). A similar argument concerning the benefits of co-location for finance and invest-ment firms suggests that these cutting-edge activities are produced in a few major cities. Co-location benefits both these sets of firms as this facilitates face-to-face interaction and the emergence of trust with potential partners, which is crucial in terms of enabling innovation and coping with the risk, complexity and speculative character of many of these activities (Sassen 1994: 84).

Both global- and world-cities analyses bring into view the wider processes shaping cities in a globalising world and economic networks amongst cities. However, the emphasis has been on a relatively small range of economic processes with a certain 'global' reach. This has limited the applicability of world-cities approaches, excluding many cities from its consideration. Although status within the world-city hierarchy has traditionally been based on a range of criteria, including national standing, location of state and interstate agencies and cultural functions, the primary determination of status in this framework is economic. as Friedmann 1995a notes, 'The economic variable is likely to be decisive for all attempts at explanation' (1995a: 317). This has become more, not less, apparent in the world-cities literature over time as more recent research has focused on identifying the transnational business connections that define the very top rank of world cities, Sassen's 'global cities' (Beaverstock et al. 1999, Sassen 2001a).

World-cities approaches have been strongly shaped by an interest in deter-mining the existence of categories of cities and identifying hierarchical relations amongst cities. This led John Friedman to ask, in his 1986 review of 'World City Research: 10 years on', whether the world-city hypothesis 'is a heuristic, a way of asking questions about cities in general, or a statement about a class of particular cities – world cities – set apart from other urban agglomerations by specifiable characteristics?' (1995b: 23). He suggests that it is both, but that the tendency has been to categorise cities into a hierarchy in which world cities are at the top of the tree of influence. This league-table approach has shaped the ways in which cities around the world have been represented – or not represented at all – within the world-cities literature. From the dizzy heights of the diagrammer, certain significant cities are identified, labelled, processed and placed in a hierarchy, with very little attentiveness to the diverse experiences of that city or even to extant literature about that place. The danger here is that out-of-date, unsuitable or unreliable data (Short et al. 1996; although see Beaverstock et al. 2000) and possibly a lack of familiarity with some of the regions being considered can lead to the production of maps that are simply inaccurate. These images of the world of important cities have been used again and again to illustrate the perspective of world-cities theorists and leave a strong impression on policy-makers, popularising the idea that moving up the hierarchy of cities is both possible and a good thing. Peter Taylor (2000: 14) notes with disapproval, though, the 'widespread reporting of [. . .] a preliminary taxonomy' of world cities. However, revised versions of these taxonomies, based on more substantial research, draw remarkably similar conclusions, and similar maps, as we will consider in the section 'Extending Global- and World-Cities Approaches', pp. 103–8 below.

In contrast to the world-city enthusiasm for categories, the global-city analysis has a strong emphasis on process. It is the locational dynamics of key sectors involved in managing the global economy that give rise to the global-city label. However, the category of global city that is identified through this subtle analysis depends on the experiences of a minor set of economic activities based in only a small part of these cities. They may constitute the more dynamic sectors of these cities' economies, but Sassen's evidence of declining location quotients for these activities in the 1990s (for example, 2001a: 134–5) suggests that the concentrated growth spurt in this sector may well be over. And it is important to put the contribution of these sectors to the wider city economy into perspective. In London, for example, where transnational finance and business services are still most dynamic and highly concentrated, the London Development Agency (LDA) suggests that only 'about 70 per cent of employment is in firms whose main market is national rather than international' (LDA 2000: 18). Even this city, routinely at the top of world-city hierarchies, is poorly served by a reduction of its complex, diverse social and economic life to the phenomenon of globalisation, and is certainly poorly described as a 'global' city. There have been many criticisms of the empirical basis for claims that global cities are significantly different from other major centres in terms of the composition of economic activities, wage levels or social conditions (Abu-Lughod 1995, Short et al. 1996, Storper 1997, Smith 2001, Buck et al. 2002).

Nonetheless, the global-city hypothesis has had a powerful discursive effect in both academic and policy circles. The pithy identification of the 'global city' as a category of cities which, it is claimed, are powerful in terms of the global economy, has had widespread appeal.[1] However, this has depended on continuing, indeed strengthening, the world-city emphasis on a limited range of economic activities with a certain global reach as defining features of the global city. This has the effect of hiding most of the activities within global cities from view, while at the same time also dropping most cities in the world from its vision. The insights of the global-city analysis are very important to understanding the way in which some aspects of some cities are functioning within the global economy. But perhaps it would be more appropriate if these processes were described as an example of an 'industrial' district. They could be called new industrial districts of transnational management and control. The core understanding about these novel processes would remain important both theoretically and in policy terms. But perhaps some of the more unfortunate consequences for cities of the global- and world-cities labels, which we will consider further below, could be avoided.

World-cities research, then, has moved on from the time of Friedmann's influential mid-1980s review of the field and, especially in the wake of Saskia Sassen's study, *The Global City*, it has adopted a strong and intensely researched empirical focus on transnational business and finance networks (see, for example, Beaverstock et al. 1999, Taylor 2004). In some ways, the focus of attention has narrowed, although there has been a concerted effort to focus on processes and to track connections amongst cities rather than simply to map city attributes (Beaverstock et al. 2000). However, cities still end up categorised in boxes or in diagrammatic maps and assigned a place in relation to a priori analytical hierarchies. A view of the world of cities emerges where some cities come to be seen as the pinnacle of achievement, setting up sometimes impossible ambitions for other cities. This also suggests to the most powerful cities that they need to emphasise those aspects of their cities that conform to the global and world cities account, with sometimes detrimental effects on other kinds of activities and on the wider social life of the city (see, for example, Markusen and Gwiasda 1994, Sites 2003). Global- and world-cities approaches expose an analytical tension between assessing the characteristics and potential of cities on the basis of the processes that matter from within their diverse dynamic social and economic worlds or on the basis of criteria determined by the external theoretical construct of the world or global economy (see also Varsanyi 2000). This is at the heart of how a world-cities approach can limit imaginations about the futures of cities and why I propose instead to think about a world of cities, all quite ordinary.

If global- and world-cities approaches offer only a limited window onto those cities that make it into the league tables, and even those at the top, we should also be concerned about the effect of these hierarchies and league tables on those cities that are quite literally off the maps of the global- and world-cities theorists. Millions of people and hundreds of cities are dropped off the map of much research in urban studies to service the very restricted view that the global and world cities analyses encourage regarding the significance or (ir)relevance of cities in relation to certain rather narrow sections of the global economy. For the purposes of developing a post-colonial urban studies relevant for a world of cities, global and world cities approaches have some serious limitations.

'FILLING IN THE VOIDS'[2]: OFF THE WORLD-CITIES MAP

In his account of cities across the world, King provocatively noted that 'all cities today are "world cities" ' (1990: 82). Unfortunately, research and writing within the rubric of the world-cities approach or hypothesis has generally not chosen to build on this observation. Because the analysis of global- and world-cities theorists has come to rely on identifying the significance of cities to only certain elements of the global economy, cities that are poorer, marginal to key globalising economic sectors or, as Manuel Castells (1983) puts it, 'structurally irrelevant' receive very little attention in this approach. However, precisely because the geography of these globalising economic sectors is changing, world-cities analysts have been drawn to explore those cities in poorer contexts that are increasingly assuming what Saskia Sassen calls global-city functions.

According to Sassen, some of the functions of command and control of the global economy also take place in what were once peripheral cities. Firms in these places are increasingly drawn to play a role in coordinating global investments, as well as financial and business services regionally. Some of this, in Latin America for example, has to do with the strong privatisation of state-owned businesses and services that has drawn more private-sector investment and has created greater demand for financial and business services (Schiffer 2002). Nonetheless, in her view, these processes signify the emergence of a new geography to the periphery – a select group of cities, some in poorer countries, are now deemed to have 'global city functions' although they fall short of being first-order global cities. She mentions Toronto, São Paulo, Mexico City, Miami and Sydney. This signals something of the 'end of the Third World' (Harris 1986) as a category in urban studies. Nonetheless, Sassen acknowledges that her approach 'cannot account for the cases of many cities that may not have experienced any of these developments' (1994: 7).

So, Sassen joins others in consigning substantial cities around the globe to the theoretical void because of their apparent structural irrelevance: 'significant parts of Africa and Latin America became unhinged from their hitherto strong ties with world markets in commodities and raw materials' (1994: 27), and, 'Alongside these new global and regional hierarchies of cities is a vast territory that has become increasingly peripheral, increasingly excluded from the major economic processes that fuel economic growth in the new global economy' (1994: 4).

Knox goes even further to suggest that 'the mega cities of the periphery will fare no better than the catatonic agrarian societies that have fuelled their (demographic) growth, and in which both will lapse decisively and irretrievably into a "slow" economic time zone' (1995: 15).

There are obviously important ways in which the changing geography of the international economy has impacted negatively on cities in poorer countries. As Sassen suggests, Western investment in poorer countries has declined precipitously since the time in the 1970s when substantial oil surpluses were recycled through poor countries as debt (1991: 83). This geography has been very uneven, though: since the 1970s Latin America has been overtaken by south-east Asia as the top destination for foreign investment in manufacturing, and many poor countries have became net exporters of capital through debt repayments (1991: 63). Nonetheless, the global-cities analysis alerts us to the role of the international financial centres of many countries in performing important 'gateway functions' for international flows of finance and the provision of global business services (1991: 173). Even so, the focus of global- and more recent world-cities work remains on a limited set of service-sector economic activities. And although global-city functions are assuming an increasingly transnational form, relatively few cities can hope to participate in them (Taylor 2004).

The 'end of the Third World' (Harris 1986) is perhaps an accurate assessment of changes over the past three to four decades in places like Hong Kong, Singapore, Taiwan, South Korea and even Malaysia, and the appearance of city-states and some major urban centres in rosters of first- and second-order global cities reflects this. But in parts of the world where global cities have not been identified – the 'voids' of world- and global-city analyses – the experience of many countries and cities in relation to the global economy has been much more uneven than global- and world-cities analyses suggest. For many, the 1980s and 1990s have been long decades of little growth and growing inequality. It is, however, inaccurate to caricature even the poorest regions as excluded from the global economy or doomed to occupy a slow zone of the world economy. Africa, frequently written off in total in these global analyses, has in fact had a very uneven growth record. As the African Development Bank (2000) notes:

> While the continent has, in overall terms, lagged behind other regions, a few countries have produced remarkable economic results, even by world standards [. . .] In an encouraging development, as many as 12 countries are estimated to have recorded real GDP growth rates above 5 per cent while close to 30 countries had positive real GDP per capita growth.
>
> (2000: 1)

It is hard to disagree that some countries and cities have lost many of the trading and investment links that characterised an earlier era of global economic relations. A country like Zambia, for example, now one of the most heavily indebted nations in the world and certainly one of the poorest, has seen the value of its primary export, copper, plummet on the world market since the 1970s. Its position within an older international division of labour is no longer economically viable, and it has yet to find a successful path for future economic growth. En route it has suffered the consequences of one of the World Bank/IMF's most ruthless Structural Adjustment Programmes (Young 1988, Clark 1989, Bonnick 1997). However, Zambia is also one of the most urbanised countries on the African continent, and its capital city Lusaka is a testimony to the modernist dreams of both the former colonial powers and the post-independence government (Hansen 1997). Today, though, with over 70 per cent of the population in Lusaka dependent on earnings from the informal sector (government bureaucrats are known to earn less than some street traders, see Moser and Holland 1997), the once bright economic and social future of this city must feel itself like a dream – albeit one which was for a time very real to many people (Ferguson 1999).

Lusaka[3] is certainly not a player in the 'major economic processes that fuel economic growth in the new global economy' (Sassen 1994: 198). But copper is still exported, as are agricultural goods and opportunities for investment as state assets are privatised. Despite the lack of foreign currency (and some-times because of it) all sorts of links and connections to the global economy persist. From the World Bank, to aid agencies, international political organisations and trade in second-hand clothing and other goods and ser-vices, Lusaka is still constituted and reproduced through its relations with other parts of the country, other cities and other parts of the region and globe (see, for example, Hansen 1994, 1997). The city continues to perform its functions of national and regional centrality in relation to political and financial services, and operates as a significant market (and occasionally pro-duction site) for goods and services from across the country and the world.

It is one thing, though, to agree that global links are changing and that power relations, inequalities and poverty shape the quality of those links. It is quite another to suggest that poor cities and countries are irrelevant to the global economy. When looked at from the point of view of these places that are allegedly 'off the map', the global economy is of enormous significance in shaping the futures and fortunes of cities around the world. For many poor, 'structurally irrelevant' cities, the significance of flows of ideas, practices and resources beyond and into the city concerned from around the world stands in stark contrast to these claims of irrelevance. As Gavin Shatkin writes about Phnom Penh, 'In order to arrive at a proper understanding of the process of urbanisation in LDCs [less developed countries], it is necessary to examine the ways in which countries interface with the global economy, as well as the social, cultural and historical legacies that each country carries into the era of globalisation' (1998: 381).

The historical legacies of these cities, it is clear from his account, are also products of earlier global encounters. Even the poorest cities have long histories of interactions and contacts with other places and have, over time, been drawn into the global economy in different roles, for trade, production, extraction or cultural exchange. These connections, perhaps transformed, can remain vital components of contemporary urban dynamics.

Viewed from off the (world-cities) map, some of the earliest versions of the world-city hypothesis are more relevant to poorer cities than later, more economistic accounts. Earlier writers suggested a range of criteria by which to assess the role and functions of different world cities. They determined whether cities were world cities with reference to economic, cultural and political processes (Friedmann and Goetz 1982, Friedmann 1995a). This means that the wider functions of many more cities can be brought into view. In these accounts, too, it is not only global processes that are relevant: the spatial reach of a city's influence is understood to vary, and there is scope for thinking about the wider role of cities in relation to their hinterland and nation, as well as to the global economy. Many more world cities are brought into view as significant provincial centres, political or symbolic centres, or perhaps as important transport and production hubs in national and regional economies (Simon 1995). Guarding against economic reductionism and moving beyond the limitations of the global scale of transnational activities would ensure that the range of cities of concern to world-cities theorists is less exclusive (Varsanyi 2000).

But there is still King's claim that 'all cities are world cities', which we need to consider, and the fact that the world-cities literature, even in its most nuanced and extensive form, persists in defining some cities out of the game, as 'excluded from global capitalism' and therefore as irrelevant to their theoretical reflections. Writers on cities in Africa, for example, asked to consider world cities in their region, conclude dismally that there are no world cities on the continent – although they point to Cairo and Johannesburg[4] as potentials (Rakodi 1997). Scholars of other peripheral places such as Latin America wonder about the usefulness of these categories in 'analysing what is occurring' (Gilbert 1998: 174), and they have been thought to have little relevance to places in the Middle East or north Africa. As Stanley writes, 'cities in this region are not on the world [cities] map' (2001: 8).

If the category of world city is not applicable to a wide range of cities (Simon 1995), are there other ways in which the world-city hypothesis might be mobilised in these 'irrelevant' cities? A stronger focus on process than categories could lead one to think about how 'global' economic processes affect all cities – as Marcuse and van Kempen (2000) frame it, this leads to a focus on 'globalising cities', since 'globalisation [. . .] is a process that affects all cities in the world, if to varying degrees and varying ways, not only those at the top of the global hierarchy' (2000: xvii). This formulation still leaves the enthusiasm about hierarchies and categories in place though and retains an emphasis on economic activities with a 'global' reach, but at least it identifies a research agenda applicable to a wider range of cities.

Most importantly, perhaps, but seldom mentioned, the particular 'global economy' that is being used as the ground and foundation for identifying both place in hierarchy and relevant social and economic processes is only one of many forms of global economic connection.[5] The criteria for global significance might well look very different were the map-makers to relocate themselves and review significant transnational networks in places such as Jakarta or Kuala Lumpur, where ties to Islamic forms of global economic and political activity might result in a very different list of powerful cities (White 1998, Allen 1999, Firman 1999, Simone 2004). Similarly, the transnational activities of agencies like the World Bank and the International Monetary Fund (IMF) who drive the circulation of knowledge and the disciplining power involved in recovering old bank and continuing bilateral and multilateral debt from the poorest countries in the world – debt, it should be pointed out, which, in an earlier phase, these agencies recommended to poor countries – would draw another crucial graph of global financial and economic connections shaping (or devastating) city life.

Global- and world-cities analyses have done much to enhance understandings of cities in a globalising economy, focusing on flows and networks amongst cities, and including some poorer cities within the same field of analysis as the most wealthy. In some ways, then, these analyses have contributed to postcolonialising urban studies, undermining inherited assumptions about the differences amongst cities. But at the same time they have reinscribed these assumptions, through a narrow focus on only a small range of global economic activities, even leaving some cities off the theoretical map altogether. Some of these points have been noted by the leading theorists of global and world cities. And there have been some new initiatives to expand the field of analysis to include more cities, and to respond to concerns about the division of the world of cities along developmentalist lines. The following section explores whether these new trends in global- and world-cities approaches can produce an analysis relevant for a world of cities.

EXTENDING GLOBAL- AND WORLD-CITIES APPROACHES?

Global- and world-cities approaches have been around for some time and have come in for a lot of criticism even as they have dominated studies of cities around the world. As criticisms have mounted, the advocates of these approaches have slowly extended their databases and entered a range of caveats into their accounts. These addenda and caveats address some of the concerns I've noted above, especially regarding the importance of paying attention to a wider range of cities than initially considered. Together, they indicate a strong movement beyond the original assumptions of global- and world-cities approaches, but also suggest the limits of this productive era in urban studies. Alternative ways of analysing cities around the world are emerging which, in my view, offer a more convincing starting point for a postcolonial urban studies committed to a cosmopolitan form of theorising. Building on the appreciation of the role of globalisation in shaping city life, accounts of 'ordinary cities', however, emphasise a diversity of economic and social networks, overlapping to produce dynamic and complex urban societies. We will return to consider the case for ordinary cities below. The path to this alternative approach, though, is paved with the criticisms and caveats acknowledged by global- and world-cities theorists themselves.

A major data-construction exercise by Beaverstock et al. has offered an opportunity to assess some of the claims of the world-cities analyses (see Beaverstock et al. 1999, 2000). Building on Sassen's identification of advanced producer services (APS) as the key distinguishing feature of the top tier of global cities, these researchers 'scavenged' (their word) data on major financial, legal, accounting, advertising and banking firms with a presence 'in at least fifteen different cities, including one or more cities in each of the prime globalization arenas: northern America (USA and Canada), Western Europe and Pacific Asia' (Taylor 2004: 65). As we noted above, this strongly skews the data set to networks of Western firms.[6] Using this data set, and differently from earlier approaches, the researchers were eager not to assume the existence of hierarchies amongst cities. Taylor notes that 'hierarchical structures are to be found in firms not in the cities' (2004: 18), and insists that whether there is a hierarchy amongst world cities must remain an empirical question (2004: 31). For him, power in networks of cities can take two forms: what he calls 'networked power', and the more familiar power to 'command and control' the global economy, which remains of interest to Sassen. As she notes, '[t]he concept of the global city introduces a far stronger emphasis on strategic components of the global economy, and hence on questions of power' (2001b: 80).

It is this concern with the relative power of cities that sees the persistent identification of hierarchies in even these nuanced responses to their critics. Cities that are very low down the ranking of connectivity to what Taylor calls the 'world city network' – in reality, a network of selected APS firms – are construed as having 'connectivity-through-subordination' (Taylor 2004: 93). While this means that even the most slightly connected global cities are acknowledged to have networked capacities in relation to APS growth, Taylor observes that 'the fact that some power has been diffused is totally swamped by [the] larger economic forces that have steered contemporary globalization' (Taylor 2004: 199). Pursuing an assessment of where power is located within world-cities networks, Taylor concludes that 'Globalization may be a world-wide phenomenon but its command centres are most certainly not so distributed' (2004: 89) and that 'globalization begins to look very "Western" as soon as direct expressions of power are investigated' (2004: 91).

It is profoundly important to remember, along with these researchers themselves, that 'Of course, these hierarchical processes are generated in the firms themselves' (Taylor 2004: 175). But they assume, or insist, that 'they carry through into an ordering of world cities in terms of high connectivities of world and regional headquarters cities' (Taylor 2004: 175). It is this slippage, from the networked power relations of firms to the ascription of a hierarchical placing of cities in relation to one another that is the source of ongoing concern about the impact of global- and world-cities analyses on the world of cities. It bears paying closer attention to the data on which these conclusions are based.

Despite a strong argument about the evidential gap within global- and world-cities research, namely the absence of evidence concerning actual flows and connections between cities as opposed to locational or attribute data from within cities (Short et al. 1996), the data that sustains the latest round of global- and world-cities analyses remains profoundly locational. The location of major producer-services firms in cities across the globe was recorded, and relations amongst different branches of these firms inferred from the relative size of different operations and the 'extra local functions' of a firm's office in a city, such as headquarters or regional offices (Taylor 2004: 66). No qualitative information was collected concerning the nature or meaning of these relationships. There is no direct evidence concerning the networking relations amongst the firms, let alone the consequences of these relationships for understanding relations amongst cities. Thus, it is not possible to draw robust conclusions about the power relations amongst cities or city status in the global economy. The evidential gap in world-cities research, unfortunately, remains.

A further problem with much of the evidence gathered in this study is its very narrow focus (Hall 2001): 'the research is very big geographically – global – but very narrow in topic' (Taylor 2004: 3). Thus it still remains focused on the very small advanced producer-services sector. Recognising this, the world-cities research team collated some evidence on other forms of globalisation, including non-governmental organisation (NGO) location in different cities, for example. Nairobi emerged top of the list of global cities defined along these criteria. A wide range of cities previously unrecognised within global- and world-cities approaches come into view in this way, which is to be welcomed. However, they still failed to capture the diverse and robust connectedness of many large cities. My own home town, Durban, for example, is labelled a near-isolate according to these criteria. Certainly it has practically no headquarters function within the national economy and it has few branches of major multinational accounting, management consulting and banking firms (although it has some). But this still misses the fact that this city of almost 3 million people is a major trading port of Africa and the second manufacturing city of a significant middle-income economy – clearly it is not an isolated place! Indeed, Durban has many different economic connections through the continent and the world and has strong connections with Johannesburg-based companies for the provision of a wide range of services, as witnessed by the very busy airlink connecting the two cities. This kind of result illustrates how a continuing investment in global- and world-cities approaches directs our attention away from the diverse and dynamic economic worlds of cities.

This example reinforces the conclusion that by relying on such narrow, location-bound indices to map world-city networks, global- and world-cities analyses cannot offer any useful assessment of the economic significance or 'worldliness' of most cities. Moreover, and quite ironically, by only measuring place-based evidence, they are unable to capture the economic importance and potential of a city's diverse connections. The evidence they do gather on a wider range of connections – including advertising, legal firms and NGOs – is quickly passed over. Taylor asserts that even though '[c]learly cities in globalization involves more than financial and business services [. . .] the latter are the dominant networkers and I continue to focus on them' (2004: 100). However, his own evidence suggests that a proper accounting for the diversity of globalisations that shape cities is now in order, including manufacturing, trade and NGOs and also informal networks. As Sassen notes in relation to the three cities (New York, London and Tokyo) she identifies as the top tier of global cities, 'Besides the vast set of activities that make up their economic base, many typical to all cities, these global cities have a particular component in their economic base [. . .] that gives them a specific role in the current phase of the world economy' (Sassen 2001a: 127). It is, I suggest, time to turn from these highly concentrated and specialist service-sector activities and pay proper attention to the vast range of activities – many also strongly shaped by globalisation – that make up the actual economies of cities.

The new data has certainly brought into view a much wider range of cities that global- and world-cities researchers are happy to call, 'global cities'. Taylor notes, for example, that there is no such thing as a non-global city (Taylor 2004: 42), as all cities have the capacity to link in to the global network when they choose to, or when they are able to mobilise the resources to do so. More strongly, he suggests that 'the world city network is not constituted as an exclusionary club of the major cities but has numerous linkages into regions beyond world cities' (2004: 78). Sassen wishes to sketch a 'decentred map of the global economy' (Sassen 2002: 30) in which cities place themselves in multiple circuits of globalisation and choose to create networked relations to neighbouring and distant places that are not central in terms of Western economies. As a result, for both writers, many more cities are now considered within the rubric of the global- and world-cities approaches. Sassen notes, for example, that in relation to producer services through the 1990s there has been 'both consolidation in fewer major centres across and within countries and a sharp growth in the number of centres that became part of the global network as countries deregulate their economies' (2001a: 118).

In the context of this wider view of global and world cities, some writers have made a stronger attempt to move beyond using the category of 'Third World city'. However, this has been replaced by a reinscription of the qualities of this category onto a certain range of cities, especially African cities. Taylor (2004) borrows from Jane Jacobs the distinction between static and dynamic cities, and draws on the analogy of entrepôt cities in colonial contexts which, he suggests, were parasitic, draining their hinterlands.[7]

However, in the contemporary period, such port cities are amongst the largest and most connected of Africa's cities and have often been the sites for diverse manufacturing development under both import-substitution and export-processing economic development regimes. Once again, paying attention to the diversity of activities and connections that shape cities would be more fruitful than simplistic assessments based on pre-given categories. It is important to recognise that an absence of outposts of large advanced producer services firms does not spell the end of the economic road or the absence of a global function for cities around the world, small or large.

Global- and world-cities approaches, however, have neglected both the wider economy and the role of local and national contexts in shaping the process of globalisation. Not only service-sector firms, but also local governments, national-policy frameworks, oppositional movements and the politics of land use all determine the potential for globalisation of city economies.[8] The complex politics of making global cities has been the focus of many other writers (for example, Abu-Lughod 1999, Hill and Kim 2000, Jessop and Sum 2000, Sites 2003). But the continuing attention paid by global- and world-cities authors to APS means that very often the policy implications of their work remain potentially detrimental to the well-being of both citizens and wider city economies. Taylor, for example, observes that there are different policy consequences for two different types of 'wannabe world cities' that he identified through the data set he was using: inner (mostly second cities in European countries) and outer (mostly 'what used to be called Third World cities' and east European cities):

> These are two distinctive policy worlds. For Outer Wannabees, rising up the ranks of world cities is primarily a 'development' issue, attracting global capital to become more central in the world city network. For Inner Wannabees, rising in world city status is about changing the nature of national city hierarchies in order to come out of the shadow of a dominant local world city.
>
> (2004: 160)

Although he and his co-researchers are eager to insist that there is 'no simple hierarchy of world cities' (2004: 165), these policy conclusions operationalise a hierarchical perspective for city competition. However, this is based on assumptions about the consequences of intra-firm relations for inter-city competition. The consequences of this analytical slippage for cities are two-fold. First, it reinforces a competitive approach to city development. Whereas cross-city collaboration may be the political strategy appropriate to firms that invest in many different cities around the world, city managers have only the well-being of their own city to consider. Attempts to move up the world-city hierarchy, if understood as the ambition of cities rather than as a result purely of the locational strategies of firms, can have severe consequences for the welfare of the residents of the city concerned. We will explore this further below. For those cities on the edges of the APS networks, Peter Hall (2001) suggests that being able to identify other forms of global connectedness that are important to the city can help to focus development strategies. But while this offers a broader development vision, it still privileges the interests and activities of those elements of the city economy that are involved in specifically global networks – a strategy which, as we will see both below and in the following chapter, can be severely damaging especially for poorer cities.

In sum, new initiatives to map world and global cities continue to entrain a concern with hierarchy, an emphasis on a small sector of the global economy, and to propose (without evidence) a close relationship between intra- and inter-firm networks and the identification of hierarchical relations amongst cities. But at the same time, both global- and world-cities advocates have ameliorated some of their earlier claims, paying closer attention to a much wider range of cities around the globe (but still exhibiting a prejudice against many important cities because of the rigid focus on APS), and entertaining the possibility of alternative forms of globalisation. However, extending global and world cities analyses in new directions has not solved some of its fundamental difficulties. In response to the critiques and caveats of world-city approaches, some other writers have thought they might well begin their analysis of cities somewhere else altogether. Amin and Graham (1997) specifically propose that we consider the 'ordinary city'. Or, as I will suggest rather more strongly, that we allow all cities to be thought of as ordinary. This move is especially important if urban scholars are to move towards a post-colonial urban theory and to generate accounts of cities and city development that are relevant to a world of cities, rather than to the limited processes that characterise 'world cities'.

THE CASE FOR ORDINARY CITIES

So far in this book I have advanced a number of arguments for ordinary cities. I have identified the importance of thinking about cities without privileging the experiences of only certain kinds of cities in our analyses; the value of learning how to think differently about cities by exploring different ways of life in other cities; and the benefits of a cosmopolitan approach to cities, including attending to the wider circulations and flows that shape them in order to appreciate the potential creativity and dynamism of all cities. A number of tactics – dislocating ethnocentric accounts, deploying comparative and cosmopolitan approaches – have been drawn on to move us towards a post-colonial form of urban theorising. At the same time, they have brought into view the ordinary city. Instead of seeing some cities as more advanced or dynamic than others, or assuming that some cities display the futures of others, or dividing cities into incommensurable groupings through hier-archising categories, I have proposed the value of seeing all cities as ordinary, part of the same field of analysis. The consequence of this is to bring into view different aspects of cities than those which are highlighted in global and world cities analyses.

First, ordinary cities can be understood as unique assemblages of wider processes – they are all distinctive, in a category of one. Of course there are differences amongst cities, but I have suggested that these are best thought of as distributed promiscuously across cities, rather than neatly allocated according to pregiven categories. And even when there are vast differences, between very wealthy and very poor cities, for example, I have suggested that scholars of these cities have much to learn from one another. This will be discussed in more detail through Chapters 5 and 6.

Second, and learning much from global- and world-cities approaches, ordinary cities exist within a world of interactions and flows. However, in place of the global- and world-cities approaches that focus on a small range of economic and political activities within the restrictive frame of the global, ordinary cities bring together a vast array of networks and circulations of varying spatial reach and assemble many different kinds of social, economic and political processes. Ordinary cities are diverse, complex and internally differentiated.

The consequences of thinking of cities as ordinary are substantial, with implications for the direction of urban policy and for our assessment of the potential futures of all sorts of different cities. Amin and Graham (1997), setting out their account of 'The Ordinary City' suggest that thinking about cities as distinctive combinations of overlapping networks of interaction leads very quickly to an account of the capacity of cities to foster creativity. In Western policy circles, they note, there has been a rediscovery of 'the powers of agglomeration', and an excitement about cities as creative centres. Agreeing that many accounts of cities highlight only certain elements of the city (finance services, information flows) or certain parts of the city – both leading to a problem of synecdoche – they rather describe (all) cities as 'the co-presence of multiple spaces, multiple times and multiple webs of relations, tying local sites, subjects and fragments into globalising networks of economic, social and cultural change. [. . .] as a set of spaces where diverse ranges of relational webs coalesce, interconnect and fragment' (Amin and Graham 1997: 417–18).

It is the overlapping networks of interaction within the city – networks that stretch beyond the physical form of the city and place it within a range of connections to other places in the world – which, for Amin and Graham (1997), are a source of potential dynamism and change. The range of potential international or transnational connections is substantial: cultural, political, urban design, urban planning, informal trading, religious influences, financial, institutional, intergovernmental and so on (Smith 2001). To the extent that it is a form of economic reductionism (and reductionism to only a small segment of economic activity) that sustains the regulating fiction of the global city, this spatialised account of the multiple webs of social relations that produce ordinary cities could help to displace some of the hierarchising and excluding effects of this approach. For with so many different processes shaping cities and so many potential interactions amongst them, it would be difficult to decide against which criteria to raise a judgement about rank.

Categorising cities and carving up the realm of urban studies has had substantial effects on how cities around the world are understood and has played a role in limiting the scope of imagination about possible futures for cities. This is as true for cities declared 'global' as for those that have fallen off the map of urban studies. The global-cities hypothesis has described cities such as New York and London as 'dual cities', with the global functions drawing in not only a highly professional and well-paid skilled labour force, but also relying on an unskilled, very poorly paid and often immigrant workforce to service the global companies (Sassen 1991, Allen and Henry 1995). These two extremes by no means capture the range of employment opportunities or social circumstances in these cities (Fainstein et al. 1992, Buck et al. 2002). It is possible that these cities, allegedly at the top of the global hierarchy, could also benefit from being imagined as 'ordinary'. The multiplicity of economic, social and cultural networks that make up these cities could then be drawn on to imagine possible paths to improving living conditions and enhancing economic growth across the whole city.

In this regard, Michael Storper (1997) has focused on the economic creativity of urban agglomerations in his description of the 'reflexive city'. Whereas global city analyses explore the importance of co-location of specialised activities involved in the control and command of the global economy in small areas of the city, Storper generalises the need for 'proximity' in economic interactions to cement relations of trust amongst complex organisations and between both individuals and organisations. Storper sees the city generally as providing a key context for these social interactions, so crucial to the 'untradeable' and 'tacit' elements of economic life. However, rather than being limited to a focus on the workings of single industry production complexes or production chains, social interaction or reflexivity is a generalised possibility in city life. He suggests then that we think of 'the economies of big cities [. . .] as sets of partially overlapping spheres of reflexive economic action [. . .] (including) their conventional and relational structures of co-ordination and coherence' (1997: 245). Cities, then, remain attractive locations for business activity across a range of sectors and offer an environment that enables economic production and innovation. This is to make a case for the broad economic potential of all cities. (Chapter 6 will explore the consequences of an ordinary cities approach for understanding urban economic development in some detail.)

Ordinary cities, then – and that means all cities – are understood to be diverse, creative, modern and distinctive with the possibility to imagine (within the not-inconsiderable constraints of contestations and uneven power relations) their own futures and distinctive forms of cityness. And there are important consequences of this approach for ways of thinking about the futures of cities. That alternative approaches to urban policy are needed is evident when the political consequences of global- and world-cities approaches are considered. These are especially onerous for the many poor cities that do not qualify for global- or world-city status. Often such cities are caught within a very limited set of views of urban development:[9] between finding a way to fit into globalisation, emulating the apparent successes of a small range of cities and, as we will see in the following chapter, embarking on developmentalist initiatives to redress poverty, maintain infrastructure and ensure basic service delivery. Neither the costly imperative to go global nor developmentalist interventions that build towards a certain vision of cityness and that focus attention on the failures of cities are very rich resources for city planners and managers who are confronted with the demands of a complex, diverse, ordinary city. Global- and world-cities approaches, once translated into a policy or political terrain, can have some serious, perhaps unforeseen consequences.

In policy terms, the hierarchies and categories embedded in the global- and world-cities approaches suggest that if cities are not to remain inconsequential, marginalised and impoverished or to trade economic growth for expansion in population, they need to aim for the top! Global City as a concept becomes a regulating fiction. It offers an authorised image of city success (so people can buy into it) that also establishes an end point of development for ambitious cities. There are demands, from Istanbul (Robins and Askoy 1996) to Bombay (Harris 1995) to be global. As Douglass (1998: 111) writes, 'world cities are the new shibboleth of global achievement for governments in Pacific Asia' (see also Douglass 2000, Olds and Yeung 2004). But, as a number of authors have noted, calculated attempts at world- or global-city formation can have devastating consequences for most people in the city, especially the poorest, in terms of service provision, equality of access and redistribution (Berner and Korff 1995, Robins and Askoy 1996, Douglass 1998, Firman 1999). Global- and world-cities approaches encourage an emphasis on promoting economic relations with a global reach and prioritising certain prominent sectors of the global economy for development and investment. Alternatively, the policy advice is for cities to assume and work towards achieving their allocated 'place' within the hierarchy of world cities (Taylor 2001).

Most cities in poorer countries would find it hard to reasonably aspire to offering a home for the global economy's command and control functions which Sassen identifies as concentrated in certain global cities. Although, as Tyner (2000) argues, different aspects of the global economy require coordination and organising and some of these activities are concentrated in cities that are not usually labelled as global. Manila, for example, has a concentration of agencies and institutions that facilitate the movement of low-paid migrant labour to wealthier countries. More feasible for many poorer cities is to focus on some of the other 'global functions' Sassen associates with global cities. These include promoting attractive 'global' tourist environments, even though these have nothing of the locational dynamics of command-and-control global-city functions. Disconnected from the concentration of arts and culture associated with employment of highly skilled professionals in global cities, the impulse to become global in purely tourist terms can place a city at the opposite end of power relations in the global economy, while substantially undermining provision of basic services to local people. (Robins and Askoy (1996) discuss this in relation to Istanbul.) In addition, export-processing zones (EPZs) may be 'global' in the sense that they are 'transnational spaces within a national territory' (Sassen 1994: 1), but they too involve placing the city concerned in a relatively powerless position within the global economy, which is unlikely to be the city's only or best option for future growth and development (Kelly 2000). These are not places from where the global economy is controlled: they are at quite the other end of the command-and-control continuum of global-city functions. More than that, the reasons for co-location would not involve being able to conduct face-to-face meetings to foster trust and co-operation in an innovative environment. Rather, they are to ensure participation in the relaxation of labour and environmental laws which are on offer in that prescribed area of the city. Cities and national governments often have to pay a high price to attract these kinds of activities to their territory. Valorising 'global' economic activities as a path to city success – often the conclusion of a policy reversioning of world-cities theory – can have adverse consequences for local economies in poor countries.

Nevertheless, global cities have become the aspiration of many cities around the world, sprawling and poor mega-cities the dangerous abyss into which they might fall should they lack the redeeming qualities of cityness found elsewhere. This may not have been the intention of urban theorists,[10] but ideas have a habit of circulating beyond our control. It is my contention that urban theory should be encouraged to search for alternative formulations of cityness which don't privilege only certain cities, placed at the top of a hierarchy. Ideas about what cities are and what they might become need to draw their inspiration from a much wider range of urban contexts.

A question that writers about cities in peripheral areas pose, looking at this theory from off the map, is how to distinguish cities they know from those that can be identified as 'world' cities. This leads quite quickly to asking how cities get to be world cities, as Alan Gilbert (1998: 178) puts it, 'So what transforms an ordinary city into a world city?' But as Mike Douglass (1998, 2000) writes, and Olds and Yeung (2004) concur, there is little explanation in this literature for 'world city formation' – or for how cities become world cities. Douglass (1998) reminds us that this is a highly contested process with profound consequences for the built environment of cities and for the well-being of citizens. Whereas the emphasis of the world-cities approach has been on understanding the 'structural' positions of cities, taking a view from off the map of this approach draws attention to political actors and institutions as active agents making the world cityness of cities (Machimura

1998, Abu-Lughod 1999, Douglass 2000, Varsanyi 2000, Lipietz 2005). These processes of world-city formation are perhaps more relevant to cities defined off the map of world cities, but eager to make their way onto it. And the processes involved in making world cities are usually not very progressive processes and are often at odds with promoting cities that are good to live in. They have been much discussed elsewhere in urban studies and include place-marketing, tourist promotion, subsidies to attract productive enterprises, costly remaking of the urban environment, all relying on often destructive forms of competition between cities and the emergence of copy-cat forms of urban entrepreneurialism (Logan and Molotch 1987, Harvey 1989, Berner and Korff 1995, Hall and Hubbard 1998, Jessop and Sum 2000, Beauregard and Pierre 2000).

Critically evaluating these world-city-making processes and incorporating them into their explanatory frameworks and empirical research – they are notably absent from the key studies within the field, for example Sassen (1994, 2001a) and Taylor (2004) – could help to sustain the critical edge of the world-cities approach and also to ensure that it remains a 'heuristic' rather than categorising device (Friedman 1995b). A greater emphasis on process rather than assigning cities to a category would certainly enable the world-cities approach to be more applicable to cities currently left off its maps. But it might also lead us to dismiss the activity of categorising cities and the category of world cities altogether. Instead, we might be encouraged to widen the range of processes considered relevant to understanding the future of cities, both geographically and functionally (Smith and Timberlake 1995). As we have seen here, the motivations for doing so are not simply intellectual accuracy or elegance; there are serious political consequences to interventions inspired by global- and world-cities analyses.

To aim to be a 'global city' in the formulaic sense may well be the ruin of most cities. Policy-makers need to be offered alternative ways of imagining cities, their distinctiveness and their possible futures. A stronger focus on the politics of urban development initiatives, as suggested by scholars of cities off the world-cities map, would expose the range of interests that find it useful to harness the global- and world-cities analyses to their ambitions. It would also bring into view the diversity of interests which are available to contest and shape the future of cities. In ordinary cities, it is this diversity – of political interests, social relations and economic activities – that can form the basis for an alternative view of cities and their futures. So, rather than develop a regulatory fiction of the powerful global city, an ordinary city perspective will start from the assumption that all cities can be thought of as diverse and distinctive with the possibility to imagine (within the constraints of contestations and uneven power relations) their own futures and their own distinctive forms of cityness.

CONCLUSION

Global- and world-cities analyses have been enormously productive in refocusing urban studies on the wider processes and networks that shape cities; and they have announced a new, more inclusive geography of the role of cities in globalisation. But they have left intact earlier assumptions about hierarchical relations amongst cities, with potentially damaging consequences especially, but not only, for poorer cities. They have, in fact, consigned a large number of cities around the world to theoretical irrelevance. Building on global- and world-cities approaches, but mindful of these criticisms, other writers have turned to the ordinary city – diverse, contested, distinctive – as a better starting point for understanding a world of cities. Ordinary cities also emerge from a post-colonial critique of urban studies and signal a new era for urban studies research characterised by a more cosmopolitan approach to understanding cityness and city futures. This can underpin a field of study that encompasses all cities and that distributes the differences amongst cities as diversity rather than as hierarchical categories. It is the ordinary city, then, that comes into view within a postcolonialised urban studies.

More than this, the overlapping and multiple networks highlighted in the ordinary city approaches can be drawn on to inspire alternative models of urban development. These would be approaches that see the potential for productive connections supporting the diverse range of economic activities with varying spatial reaches that come together in cities. Approaches that explore the diversity of economic activities present in any (ordinary) city (Jacobs 1965: 180–81) and that emphasise the general creative potential of all cities could help to counter those that encourage policy-makers to support one (global) sector to the detriment of others.

The following two chapters explore some of the arguments and foundations for development interventions and theoretical perspectives that might be more appropriate to ordinary cities with diverse economies. Chapter 5 turns to a body of research and policy-making inspired by the problem of developing poorer cities. In some ways, the mirror image of global- and world-cities approaches, developmentalist approaches to cities inspire an emphasis on the poorest parts of cities, those lacking in much of the basic infrastructure taken for granted in wealthy urban contexts and those with little in the way of cutting-edge formal global business. To develop a post-colonial account of ordinary cities, the divide between global- and world-city analyses and developmentalist urban approaches needs to be breached; the diversity and complexity of ordinary cities needs to be brought into view.

Both of the remaining chapters in this book explore how scholars working on poorer and wealthier cities have much to learn from each other. I explicitly set out to inspire a more cosmopolitan form of theorising: one that tracks across different kinds of cities and that, like the inventive and innovative urbanisms explored in Chapter 3, learns promiscuously from a range of contexts. Thus I also insist that imagining the futures of ordinary cities requires more than a developmentalist approach, just as it requires more than a global- or world-cities approach. These currently divided areas of urban scholarship need to be brought together, in both theory and policy, if urban studies is to find the resources to address the challenges of an urban future; the challenges of a world of (ordinary) cities.

5 Bringing the city back in

Beyond developmentalism and globalisation

INTRODUCTION

Where global- and world-cities approaches have emphasised those cities and parts of cities where 'command and control' functions of the global economy are located, developmentalist approaches to the city have tended to focus on those parts of cities that are far less successful, with many different demands for basic services, infrastructure and employment. Neither of these approaches is necessarily inaccurate, but each brings only a limited part of the city into view. In contrast, urban-development practitioners have become increasingly concerned with thinking about the city as a whole and with taking action 'at city level' (DFID 2001: 31; also World Bank 2000: 62) in order to be able to address the many different challenges posed by cities, especially resource-poor cities. Partly inspired by this trend within development policy, this chapter makes the case for thinking through the diversity and complexity of the city – bringing the city itself back into view within urban studies. The chapter suggests that there are also good analytical reasons for looking not only at the globalising networks that shape cities, as we saw in the previous chapter, but also at the dynamic spaces of cities themselves.

The rich complexity and diversity of city life presents a significant challenge to city managers, citizens and development organisations. Many cities around the world encompass both rapidly globalising economies and infrastructurally poor neighbourhoods. And these coexist alongside a range of diverse enterprises, informal activities and different neighbourhood contexts, many neither flashily global nor desperately poor. Facing up to this complex, contested and diverse urban context, policy makers, local authorities and communities working in particular cities have many different, often competing, agendas to address; they have had to invent urban development policies appropriate for ordinary cities.

For urban studies to contribute to development strategies for ordinary cities, it will need to offer analyses that have a purchase on the diversity of economic activities, political interests and the range of needs of citizens. Rather than theories of world cities, which encourage competitive behaviour amongst cities and focus on a very small segment of the urban economy, or developmentalist accounts, which zoom in on the poorest and least well-provisioned areas of the city, urban studies will need to explore the dynamics of socially and economically diverse cities. Both globalising economic sectors and the needs of the poorest citizens will have to be considered, bringing the ordinary city, in all its complexity and diversity, firmly into view. And it is the argument of this chapter and the next that bringing the city back into view is a prerequisite for framing more inclusive and redistributive urban policies.

In turning to learn from debates within urban development studies, this chapter offers yet another reason to abandon the divide within urban studies between studies of wealthier and poorer cities. Learning from analyses of cities in poorer contexts mobilises more cosmopolitan resources for a post-colonial urban theory. This facilitates an appreciation of the diversity of cities and has the potential to inform analyses relevant to a world of cities.

The first section of this chapter, 'Bringing the city into view', draws together arguments for bringing the city back in to urban studies. Rather than limiting enquiries to global networks of information and power (Castells 1996, Taylor 2004), or to small localised parts of cities (Sassen 2001a), this section asserts the value of city-wide analyses. The spatialities of cities, it suggests, are multiple and accounts of cities would benefit from being thought of as not only bundles of flows and networks, but also, for example, as distinctive territories, arenas of contestation and platforms for growth. In this regard it is suggested that there is much to learn from the field of urban development studies, where discourses on urban development and interventions in poorer cities have increasingly scaled up, shifting from an earlier emphasis on localised projects and the poorest neighbourhoods to promote inclusive development and policy formulation across the city. Most recently, these initiatives have been formalised as 'city development strategies' (CDSs) and have been promoted by a range of international development agencies. They are discussed in the third section of this chapter, pp. 126–132 below, after which the chapter turns to the example of the city of Johannesburg, where a version of a CDS was prepared between 1999 and 2002. Confronting this unequal and complex city in its entirety for the first time in the post-apartheid era, policy-makers and scholars here have had to creatively move beyond both global city analyses and developmentalist prescriptions. This has involved responding to a divided and contested social and political context, as well as engaging with a complex and diverse economy. A central point here is that the process of bringing the city into view within both policy and theory is fraught with political conflict. Nonetheless, this case suggests that all cities would benefit from being understood as ordinary, inviting analyses and interventions that stretch across the range of diverse activities and interests that they bring together.

Bringing the city back in provokes new questions for cities – about linking economic growth and basic service delivery, for example, or addressing the needs of the poor alongside facilitating the activities of globalising firms, building connections between formal and informal businesses, linking formal and informal institutions of governance or thinking about the generalised advantages of cities as locations for diverse kinds of economic activity as well as specialised clusters of innovative enterprises. Letting cities be ordinary means thinking across the many different elements of city life and across a diverse range of cities. This chapter and the next bring out the implications of this approach for both theoretical analysis and policy debates.

BRINGING THE CITY INTO VIEW

It is paradoxical – so many writers argue – that in an era of globalisation the importance of place for some social and economic processes may have been heightened. Classically, and precisely because of the transnationalisation of many business activities, the growing significance of advanced producer services and financial services has apparently led to a resurgence in demand for central locations in major cities.[1] Some firms apparently prefer to be located near one another, and the importance of proximity, or co-location, in certain parts of cities has been much explored (Boden and Molotch 1994, Storper and Venables 2004). This is certainly not true of all economic activities nor of all cities, but the phenomenon of clusters of certain kinds of firms in some cities has driven a strong sense of the continued importance of localisation even in a globalising and strongly interconnected economy. By all accounts, place continues to matter in familiar ways – the benefits of co-location for interaction, face-to-face meetings, access to common facilities – and in new ways, to manage and coordinate certain complex transnational economic activities, for example.

But the places in question are usually very small areas of cities. Clusters, whether specialised in one sector or multi-sector clusters of diverse firms, are often thought of as geographically concentrated, perhaps in a small part of a city (Sassen 1991), or have consolidated over many years in a particular neighbourhood (Benjamin 2000). They usually only bring into view small segments of the city. For writers considering the continuing significance of place for economic activities in global and world cities, the city as a whole does not seem very relevant. It is as if the location of these activities within the city – rather than the neighbourhood – were not important: the rest of the city, its political dynamics and its range of economic activities, is left out of the equation.

A different spatial imagination is necessary to view the ordinary city. Contemporary analyses of the territorialisation of the global political economy (Harvey 1989, Jessop 1994, Brenner 1998), suggest that the significance of city-level government has expanded. The 'hollowing out' of the national state – or at least the reconfiguration of its responsibilities – has arguably happened to the benefit of city governments. Other processes have also been at work to territorialise political power at the city scale. In many post-socialist, post-authoritarian and poorer country contexts, processes of democratisation have seen decentralisation of political responsibilities to cities. These changes have taken place within a globalising economic context of footloose manufacturing capital, circulating financial capital and a globalising service sector. Newly empowered cities across the world have joined a growing number of entrepreneurial, relatively autonomous local governments concerned to promote and to expand economic activities in their cities.

However, the role of local political processes in shaping cities is only reconfigured under globalisation; it is not new: there is a long history of proactive political engagement in making city spaces. Processes of sectoral concentration and economic inequality, often ascribed to the late-twentieth-century globalisation of finance and service-sector activities, are also shaped by much longer, local economic histories. As Abu-Lughod concludes in her comparative analysis of New York, Chicago and Los Angeles, 'common forces operating at the level of the global economy operate always through local political structures and interact with inherited spatial forms' (1999: 417). The specific (strongly racialised) local politics of housing, education and planning intersected with national economic and policy trends to frame divergent, but not unrelated, trajectories for these most international cities of the USA. Features often ascribed to the recent concentration of global-city functions Abu-Lughod tracks through deep historical analysis. Not only are these global-city functions of much longer duration than usually acknowledged; their apparent consequences, such as income inequality, dual labour markets and informalisation of immigrant labour have long histories too and can be ascribed as much to taxation changes, racism, migration policy and declining local-welfare nets as to the external processes of economic globalisation. In addition to the wider processes of globalisation, the complex political spaces of the city also determine specific urban trajectories and in turn shape the dynamics of wider economic processes (Kelly 2000, Jessop and Sum 2000).

With this in mind, Scott et al. propose that cities are playing an important role in supporting and enabling new trends in economic globalisation. And they suggest that it is at the widest scale of the city-region that these processes are important. They observe that:

> rather than being dissolved away as social and geographic objects by processes of globalization, city-regions are becoming increasingly central to modern life, and all the more so because globalization (in combination with various technological shifts) has reactivated their significance as bases of all forms of productive activity, no matter whether in manufacturing or services, in high-technology or low-technology sectors. As these changes have begun to run their course, it has become increasingly apparent that the city in the narrow sense is less an appropriate or viable unit of local social organization than city regions or regional networks of cities. [. . .] Large city-regions are thus coming to function as territorial

platforms from which concentrated groups or networks of firms contest global markets.

(2001: 11, 14)

In their view, large cities, including those in poorer or 'developing' regions,[2] offer the opportunity to respond readily to innovations and changing conditions by drawing on the wide range of services, activities and concentrated expertise in large urbanised areas as well as on the specialised interactions enabled by concentrations of related firms in the city (Rodriguez-Posé et al. 2001, Rogerson and Rogerson 1999, Buck et al. 2002). This is especially the case for firms operating in sectors of relative uncertainty and rapid change. With urban sprawl, edge cities, suburban and exurban business parks and the emergence of strongly interconnected systems of cities in some parts of the world, Scott et al. suggest that rather than simply cities, it is 'city-regions [that] now increasingly emerge as privileged sites of generalized competitive advantage' (2001: 21).

Large cities, which encompass a wide range of services and many different functions and activities, offer a generic platform for diverse economic activities and the opportunity for experimental and innovative ventures (Duranton and Puga 2000). They also offer a variety of locations for economic activities – from concentrated, dense city centres to disused and cheap warehouse districts or manicured, state-of-the-art business parks in secure suburban locations. Opportunities within broader city-regions for the decentralisation and even reconcentration of activities in nearby cities suggest that it is the dynamic and uneven territory of the extended city that functions as a platform for economic activity.[3]

However, the proposal that it is the wider city-region, rather than just the 'city', that offers the territorial context for much contemporary economic activity unsettles any easy delimitation of the spatial extent of cities. But even cities whose boundaries were defined by physical walls were maintained through flows and connections with hinterlands, with other cities, with distant parts of the world (Taylor 2004). The difficulties of determining the territorial extent of cities, while perhaps more acute in an era of globalisation and sprawling city-regions, has been a perennial of urban sociology and urban politics.[4] Clearly cities cannot be understood as territories in any sense of being firmly bounded, easily demarcated or contained. But the complexity of a city's social and political life, the diversity of economic activities and spaces and the multiplicity of flows and networks that operate in and through cities constitute them as distinctive places, as sites for social, political and economic activity. As Le Galès (2002) observes of European cities:

Any study of cities must steer a course between the Scylla of representing the city as a separate unit, thus risking its reification, and the Charybdis of showing it to be infinitely diverse and complex. Avoiding the first of these risks means taking into account the diversity of actors, groups, and institutions that make up the city. The city is also by its very nature fluid, confused, full of movement, and made up of individuals who love, work, have fun, trade, and participate. [...] The study of local actors, interactions, and diversity is not enough on its own: it must be accompanied by research into the mechanisms for integration and the modes of cooperation that help to construct an urban social and political order, fragile and ephemeral though this may be.

(2002: 183, 185)

The site, or place, of the city might be constituted, as Le Galès suggests for European cities, through practices of political cooperation and long histories of communal relationships. It is also shaped through practices of government; the creation of political communities that stretch across the diverse social, political and economic relations of any city or city-region can constitute a powerful, if usually profoundly contested, space of redistribution and regulation and provide opportunities for the formulation of common trajectories. This might involve looser confederations of smaller authorities and relate only to single-function regulators, such as transit authorities (for New York, see Abu-Lughod 1999, Sites 2003), or water-catchment management (for Brazil, see Abers 2004), or involve responses to social inequalities and metropolitan disintegration (see Brenner 2002, Savitch and Vogel 2004). Large cities and city-regions are often able to develop close political and economic interrelations despite fragmentation and contestation. Even dispersed and diverse urban regions can cohere, perhaps only temporarily and ephemerally, around collective identities. Cities such as Los Angeles, São Paolo and London stretch across many different local authorities and neighbourhoods and encompass deeply contested political and cultural identities. But the identity of belonging to a distinctive place remains current and available, and may well draw precisely on the characteristic social and political relations across the city (such as the distinctive practices of cosmopolitanism in the case of London, or racialised division and insecurity in the case of Johannesburg), to constitute such place identities.

The territorialisation of economic activities, political relations and place-based social identities offer opportunities to engage with the city as both a place (a site or territory) and as a series of unbounded, relatively disconnected and dispersed, perhaps sprawling and differentiated activities, made in and through many different kinds of networks stretching far beyond the physical extent of the city. The spatial imagination adequate to capturing cityness – in its diverse forms – must necessarily be multiple and sophisticated. Networks and localised clusters; boundaries and globalising flows; communities of responsibility as well as divisive social fragmentation: cities are all these and more.[5] To lose or occlude some of these aspects of urban spatiality can have serious consequences – as I argued, for example, in relation to the prominence given to globalising networks and small areas of the city in global-cities approaches. In motivating to bring the city back in to urban analysis, I am by no means suggesting that the spatialities of networks, flows, fragmentation and diversity are less relevant. In fact, I would suggest that it is precisely by paying attention to the city – as a territory, a platform, or in relation to city-wide processes – that the diversity of city life and the multiplicity of networks and connections that shape it come back into view.

The diversity, fragmentation and complexity of cities means that although cities can usefully be thought of as places, or territories, it is not possible to capture the entirety of any city: the city 'as a whole' can never be known (Pile and Thrift 2000). Historically, plans to determine the form of the city, to control urban social life or to influence the direction of urban change on the basis of an all-encompassing view of the city have fallen foul of the complexity and unpredictability of cities (de Certeau 1984). And, certainly, it is not possible to hope to represent all the activities, or all the concerns of people, or even all interest groups, in a city. What is emerging, though, in urban development policy circles – as the following section explores – is an openness to engaging across the diversity of the city. I want to argue here that taking a city-wide view, alongside a proper appreciation of the complexity and diversity of cities, can do some important work in reconfiguring urban theory, moving beyond a Western/Third World city divide and beyond the respective limitations of global-cities or developmentalist approaches. In this regard, it is useful to turn to consider some recent changes in urban-development policy, where the value of bringing the city back in, in all its diversity and complexity, has come to be appreciated.

DEVELOPMENT AND CITIES

Partly because scholars of so-called 'structurally irrelevant', or 'Third World' cities find it hard to pursue research and policy within the frame of world-cities analysis,[6] urban development studies has maintained a distinctive research agenda while global- and world-cities analyses have risen to prominence. As Browder and Godfrey (1997) point out, '(b)eyond general inferences, the implications of recent world city formation for Third World Urbanisation go largely uncontemplated' (1997: 45). Instead, developmentalist analyses have offered an alternative frame of reference for many cities in poor countries.[7]

A substantial literature has developed on various aspects of urban development: community participation, housing, land tenure, service provision, governance capacities, infrastructure, informal sector, and so on. All of these are crucial in so many ways for improving the living environments and livelihoods of people in poor cities. But they very seldom show up on the radar of the wider field of urban studies, which claims to be concerned with the dynamism and centrality of urban life in the contemporary global economy. The one place where some of these concerns about immense urban poverty do emerge into a wider urban studies is in the considerations of mega-cities. Big but not powerful (Beaverstock et al. 1999), mega-cities attract other forms of theoretical fascination, with the more difficult and adverse consequences of urbanisation (Beall 2000, Lo and Yeung 1998).

However, in the same way that global- and world-cities approaches ascribe the characteristics of only parts of cities to the whole city through the process of categorisation, mega-city and developmentalist approaches extend to the entire city the characterisation of those parts that are lacking in all sorts of facilities and services. Where the global-city approach generalises the successful locales of high-finance and corporate city life, the developmentalist approach tends towards a vision of all poor cities as infrastructurally poor and economically stagnant yet (perversely?) expanding in size. Many other aspects of city life in these places are obscured, especially dynamic economic activities, popular culture, innovations in urban governance and the creative production of diverse forms of urbanism – all potentially valuable resources in the quest for improving urban life (Rakodi 1997, Askew and Logan 1994, Hansen 1997, Simone 2001). Envisioning city futures on the basis of these partial accounts is certainly limiting. Moreover, from the point of view of urban theory, developmentalist accounts do not necessarily contribute to expanding the definition of cityness; in fact they are frequently drawn on to signify its obverse, what cities are not. Consequently, there have been long-standing criticisms of the category of Third World cities: that urban diverse experiences are unhelpfully pulled through the common lens of 'Third-World-ness', and that the distinctive features of these cities are identified and understood primarily in relation to normative Western experiences.[8]

Nonetheless, the split in urban studies identified here has been reinforced by the rise of 'Third World'-ism and the field of development studies, specifically concerned with speeding up the economic growth of less developed countries (Hewitt 2000). Within this framework, the poorest cities in the world have been characterised by their distinctive features as 'Third World cities'. Although hampered by the idea of urban bias in which cities were seen to be draining the countryside economically, over time a set of strategies have evolved that are designed to help cities in 'Third World' countries address what seem to be their very different concerns from cities in the West – rapid population growth without economic growth; burgeoning informal sector activities; a large poorly housed or homeless population and extensive irregular settlements (Rakodi 2002).

While cities have been seen as distinctive sites for interventions in the form of targeted development projects mostly at the neighbourhood level, until very recently the city as such has been considered broadly irrelevant to development. More specifically, until recently there has been very little attention in policy circles to promoting cities as particular sites of economic growth. Urban economies have conventionally been seen as the outcome of national and international decisions and considered to be the province of authorities at these scales. National development strategies such as import-substitution industrialisation had substantial consequences for urban growth, as manufacturing firms and infrastructural development transformed cities and provided employment opportunities for the growing population. But the field of 'urban development' had neglected what Nigel Harris (1992) has called the 'real urban economy' for some decades. Economic growth was not considered an important part of urban policy and was much more the province of national and regional level governments.

Policy-makers interested in promoting development have since come to appreciate the significance of urban economies, whose successful management and development is now seen as a crucial determinant of wider economic growth. Starting with a major World Bank (1991) initiative, this approach saw cities as 'engines of economic growth' rather than parasitic drains on the national economy. They emphasised enabling and partnership strategies for housing and services provision (as opposed to state or donor provision) and highlighted the importance of infrastructure provision and efficient city-wide managerial capacity as essential to support economic enterprise.

From the point of view of addressing poverty, too, the stretching of the urban-development imagination to include the city as a whole, rather than targeted projects (although these remain important forms of development intervention), is increasingly seen as key (Acioly 2001, Loughhead and Rakodi 2002). The UK's Department for International Development in India explain their changing focus in the following way:

> DfID's work in urban poverty illustrates how its programme has changed over the years. From the early 1990s, DfID was involved in physical slum upgrading projects. This had visible and tangible outputs, but raised questions about how far the poorest and most vulnerable benefited. Since the mid-1990s our projects have looked more at improving services to the poor through participatory planning and attempts to link slums into the city systems. By the end of the 1990s it was realised that institutional factors, such as improved city management and more responsible local government, are pre-requisites to improved services for the poor.[9]

Furthermore, as the 2001 Global Report on Human Settlements notes, addressing inequality is as effective a way of combating poverty as promoting economic growth, if not more so (UNCHS 2001: xxxii–xxxiii). But for that both poor and wealthy parts of the city need to be considered together. From this perspective it is imperative that the imaginations of the world-city analysts and developmentalist urban policy are drawn together.

Recently, then, there has been a theoretical convergence of sorts as advocates of urban economic development policies have turned to analyses of globalisation for inspiration (Harris 1992, 1995, Cohen 1997). Partly as a result of the World Bank's earlier policies, the role of city government is now understood to include the promotion of urban economic development. As Harris summarises:

hitherto 'urban development' has tended to exclude a concern for the underlying urban economy, making it impossible for city authorities to consider directly measures to enhance urban productivity. The agenda has been broadened from the immediate issues of maintaining order and providing services, to a concern with the environment of the poor. It needs now to consider the economy proper, particularly because increased administrative decentralisation and a more open world economy are likely to make the role of city managers much more important (however these are identified). This will require considerable inputs of technical assistance, particularly to identify the city-specific agenda of issues and continuing mechanisms to monitor the changing economy. Hitherto, local authorities have had little incentive to trouble themselves about the economy within their administration. However, decentralisation with greater democracy could enforce on local authorities an increasing interest in the sources of the city's revenues as well as the citizens' income.

(Harris 1992: 195)

Urban development initiatives at the end of the 1990s have dovetailed with substantial administrative decentralisation in poorer countries to produce a set of policy proposals focusing on promoting urban economic development at a local level. These initiatives are also reinforced by a growing awareness of the competitive role of cities across the world in the 'global' economy (Wolfensohn 1999, Stren 2001). Drawing on and extending the experiences of local economic development initiatives already prominent in many Western cities, urban development policies in poor countries at the turn of the century have started to follow the path that Harris was predicting at the beginning of the 1990s. Urban-development initiatives at a city-wide level (CDSs) are advocated by major international agencies and increasingly implemented by cities around the world (Campbell 1999, World Bank 2000, UNCHS 2001). We will explore these in more detail below.

Within the developmentalist framework, then, cities have come to be understood as significant new territorialisations of the global economy. This matches the analyses of writers such as Brenner (1998) in some ways, although the reasons for the expanded role of cities in the global economy are somewhat different in these cities than in Europe, for example. Here it is administrative decentralisation, democratisation, tighter aid and policy control by international financial institutions (IFIs), as well as new forms of economic liberalisation that have contributed to the growing recognition within development practice of the city as a significant site of developmental planning (Robinson 2002b). Increasingly, policy-makers suggest that cities that are well organised and managed can build on their own distinctive combinations of economic activities and broader assets to act as a competitive platform for attracting and directing economic investment and encouraging economic growth. This way of thinking about cities and their potential for development has much in common with other prominent approaches to local economic development under neo-liberalism and global-isation,[10] but is sensitive to the social and economic diversity of cities and doesn't assume their subordination to any particular global logic.

In poorer cities it is precisely the coexistence of local and translocal, formal and informal economic activities, in the context of a desperate need for basic services, that is challenging policy-makers[11]. Both global- and world-city approaches and developmentalist policy frameworks have little to contribute on how to work with the diversity of these cities, rather than only charac-teristic segments of them. Neither of them offer us many resources for imagining possible development paths that cut across and work with the coexistence of 'global' formal activities and translocal informal trading or help foster links between city-wide or neighbourhood-based firms as well as the transnational firms that sustain many economies. The challenge for urban studies, then, is to develop creative ways of thinking about connections across the diversity and complexity of city economies and city life. This is not simply to more accurately represent and understand cities, but to contribute to framing policy alternatives that can encourage interventions in a variety of social and physical urban environments and can also support a diversity of economic activities with a wide range of spatial reaches, rather than prioritising only those with a global reach. This could also ensure that urban interventions address the inequalities that stretch across and between cities and that sustain poverty in them (Beall 2000, UNCHS 2001). For approaches to urban development that are both inclusive and redistributive, then, it is important to bring the city back in to urban studies.

CITY DEVELOPMENT STRATEGIES

If the city as such has fallen out of view in much urban theory, then international donor agencies increasingly have them in their sights. City development strategies (CDSs) have come to be supported by international development agencies, especially the World Bank and the United Nations Centre for Human Settlements (UNCHS) in the hopes of bringing pro-growth and anti-poverty agendas together. By December 2002 Nigel Harris could report that 'well over 100 CDSs must have been undertaken' (2002: 2). The CDS envisages the formation of local stakeholder forums, where a vision and a strategy for improving the city can be developed on the basis of a 'socio-economic map of a city' (Cities Alliance 2002), commonly prepared by consultants or city officials. The city vision would aim to incorporate the diverse concerns and needs of citizens, businesses and local government. This could mean attending to the impacts of globalising sectors of the economy alongside the needs of the poorest citizens, as well as appreciating the wide range of activities that contributes to the dynamism of cities. In contrast to global- and world-cities analyses, and also quite differently from earlier development interventions in cities, CDSs build an approach to cities that requires a city-wide view and engages with the complexity and diversity of the city. Although they certainly have some limitations, both in principle and in practice, they offer an important example of the consequences for urban-development policies of seeing cities as 'ordinary', and are an attempt to face up to the challenges presented by the diversity of city economies and societies.

Since the late 1990s, urban development agendas in poor countries have been more strongly targeted by national governments and international agencies of various kinds.[12] While these initiatives draw attention to the diversity of the city and make a potentially important contribution to understandings of cities more generally, a significant political consequence of these new initiatives is the globalisation of local government itself. The politics of information circulation, international networking and conferences, advice and technical support, the prominence of donor agendas and consultants' analyses, as well as the effects of powerful discourses about urban growth and development: all these have the potential to limit local decision-making and impose external agendas. So, even as more limited developmentalist urban agendas are being opened up to engage with the economic and social diversity of cities through the promotion of CDSs, local governments and the futures of cities in poor countries are increasingly enmeshed in the international ambitions and political discourses of wealthier countries, and the analytical approaches of development specialists and urban consultants. Central to CDSs then, is that they have emerged within a field of developmental practice in which the focus is very much on cities in poor countries, so it is important to bear in mind that the power relations that shape the field of development practice remain relevant to understanding their likely impact (see Escobar 1995, Watts 2003). This makes it especially important for urban studies scholars concerned for redistributive urban policies to engage with the intellectual terrain in which policy advice is formulated.

CDSs are in part a response to significant changes in the wider policy context of local government and the management of poorer cities. As estimates of the proportion of people in poorer countries who live in cities has risen, international development agencies have had to pay more attention to urban issues in development (see Table 5.1).

Table 5.1 Percentage of population residing in urban areas by region (estimated in 2003)

	1950	1970	1990	2010	2030
World	29.1	36.0	43.2	51.3	60.8
More developed regions	52.5	64.7	71.8	76.1	81.7
Less developed regions	17.9	23.7	35.2	45.9	57.1
Least developed countries	7.4	11.1	20.9	30.4	43.3

Source: United Nations Department of Economic and Social Affairs, Population Division, 2004: 30–1.

Of course the overall proportion of the urbanised population in designated 'less developed areas' varies widely across countries and regions and is very dependent on the different definitions of what counts as urban in each country. But the headline statistics of an urbanising world population, as well as the growth of very large cities in the poorest countries of the world have been influential in shaping policy-makers' perceptions of priorities and are repeatedly cited in urban development policy documents and related literature. The World Bank opens its strategic policy document initiating a stronger urban focus for the 2000s as follows:

> At the threshold of the 21st century cities and towns form the frontline in the development campaign. Within a generation the majority of the developing world's population will live in urban areas and the number of urban residents in developing countries will double, increasing by over 2 billion inhabitants. The scale of this urbanization is unprecendented and poses daunting requirements for countries to meet the needs of their people at relatively low levels of national income.
>
> (2000: 1)

The same document closes with the rousing statement that 'The future is urban, and we must be there' (2000: 79). It is particularly the association between high levels of urbanisation and increasing poverty that has motivated action on the part of development agencies. For the President of the World Bank, James Wolfensohn, '(t)he fight against poverty in global city-regions [. . .] is truly a fight for peace and stability on our planet' (2001: 49). Given the already high levels of urban residence in wealthier countries, offering little scope for further urbanisation, and the very large populations and high rates of urbanisation in some of the poorest countries, such as China and India, the assessment that for the foreseeable future poverty is likely to be closely related to urbanisation is widely supported (see, for example, Hall and Pfeifer 2000, UNCHS 2001, Montgomery et al. 2004). These observations have prompted a shift in the emphasis of policy-makers and practitioners towards a much stronger urban focus for development interventions.

In addition to the impact of these population projections, changing understandings of the place of cities in the global economy have played a very important role in reconfiguring urban development policy (see Stren 2001). The sense that capital is relatively footloose and able to move between competing cities frames a concern for the volatility of urban economies and a desire to support active management of cities to ensure they can maintain current investment and encourage future growth without increasing inequality (UNCHS 2001). Added to this is the assessment that many cities offer services (financial and business services especially) and support for a wide range of economic activities, and are therefore key sites for the location of globalising business activities and potentially 'strategic nodes in international networks of exchange' (Montgomery et al. 2004: 24). As DfID note in their urban strategy document, 'the impact of globalisation, [. . .] has increased the flow of goods, services, capital and resources between regions and urban centres, and thus shifted attention to the efficiency and effectiveness of different city management systems to encourage inward investment' (2001: 22).

So, the world is becoming more urban and city economies are becoming more globally connected: both good reasons to bring urban development more firmly onto the agenda. But also important in the shift to emphasise urban development are the decentralisation of political responsibility and the promotion of commercially driven forms of urban management. Both of these have contributed to the restructuring of local government activities in many parts of the world (Campbell 1997). Apart from increased responsibility for service delivery in urban areas and injunctions to build partnerships, encourage participation and develop new kinds of funding structures, in the context of a globalising economy local-government mandates now commonly incorporate the promotion of economic growth as well (Amis 2002, Harris 2002).

Securing growth at the same time as expanding service delivery in politically contested and resource-poor environments represents a great challenge for local governments. Electoral or popular support may be consequent upon developing effective services and, increasingly, ensuring that private firms meet the needs of the poor; on the other hand, long-term viability or national state approval may depend on promoting dynamic economic growth.[13] And, in many circumstances, local government comes a very distant second to more informal measures for achieving all these aspects of urban governance, as service delivery, economic expansion and political authority are produced outside of formal state structures (Halfani 1996, Simone 2004). Forging effective interfaces between local institutions of government and informal urban processes represents yet another challenge of governance in many cities (World Bank 2000: 7).

CDSs, then, are a policy response on the part of international development agencies to these diverse and, at times, contradictory new demands on city government. According to the Cities Alliance (a broad association of international agencies, associations of local governments and national governments concerned with urban development) CDSs, which along with slum upgrading constitute one of their two strategic initiatives, involve individual cities developing 'a collective city vision and action plan aimed at improving urban governance and management, increasing investment to expand employment and services, and systematic and sustained reductions in urban poverty' (Cities Alliance 2002). For the World Bank, a core member of the Cities Alliance, the CDS involves stakeholders identifying ways to achieve urban sustainability by aiming to produce liveable, competitive, well-governed and bankable cities. This establishes a wide-ranging agenda for addressing poverty, slums, environmental degradation, economic growth, service provisioning and institutional capacity.

Although reducing poverty is one of the 'corporate goals' of the World Bank (2000: 40), the policies of individual donor countries and the UNCHS express a much stronger commitment to drawing on city-wide development strategies to deliver development that meets the needs of the poorest urban residents. DfID, for example, sees a CDS as 'an effective tool for developing a new approach to stimulating economic development, improving urban governance and management approaches, in order to bring about a sustainable reduction in urban poverty' (2001: 31). In all discussions of CDSs, the construction of a broad-based participatory structure where a collective city vision can be forged by a range of different 'stakeholders' and 'owned' locally plays an important role. Often, within a developmentalist framework, it is effective participation by the poor and community organisations that is seen as the key to delivering an inclusive development agenda for a city (GHK Consultants 2000).

In principle, CDSs offer a conceptual and organisational space for considering the diversity of activities, concerns and needs in cities. Whereas making the city a suitably attractive and competitive base for international investment would probably be an important part of any CDS, the commitment of agencies as diverse as the World Bank and UNCHS is to fostering an inclusive and participatory approach to city-visioning. Supporting informal sector activities, addressing service delivery needs, encouraging pro-poor policy initiatives and institutional capacity building are all imagined as within the ambit of a CDS, but so are initiatives to reform service delivery mechanisms (privatisation) and to facilitate access to financial resources for major projects and programmes. Here, then, is a final motivation for promoting CDSs, for the World Bank at least, which is that cities are 'good business for the Bank' (World Bank 2000: 4, 40). The bank imperative to generate profitable lending relationships with clients and to access grant income for relationships involving non-lending technical advice makes the formalisation of strategies for urban development at least in part a self-interested initiative, although the World Bank suggests that it faces growing demands for support from newly empowered municipalities (2000: 13, 61).

Other agencies also see CDSs as a way of delivering on their own ambitions whether this be the eradication of poverty, promotion of a 'pro-poor' development agenda or support for progressive and redistributive responses to globalisation in cities (DfID 2001, UNCHS 2001). The authors of the DfID urban strategy document note, for example, that 'actual stakeholder involvement [in CDSs] is often being driven by the necessity to tackle ever mounting urban service delivery problems, rather than to building community wide capacities and capabilities' (2001: 32, also GHK Consultants 2002). Developing broad pro-poor policies are not always the priority for the poor, or for local governments, and donor priorities have the potential to overshadow locally negotiated visions, especially in the poorest cities and especially where donors offer scarce resources in support of their own agendas.

A range of informal interactions, networking relationships and synergies amongst the urban agendas of multilateral and national development agencies suggest, then, that a new architecture of international policy in relation to city development is being forged. On the one hand, this productively supports innovative urban-based research and policy, which has been relatively neglected within development interventions; on the other it has the potential to cement the power of wealthier countries and technical experts to shape the development paths of cities in poorer countries. New networking opportunities, such as the annual Competitive Cities Conference, cut across conventional Western/Third World city divides, for example, but the US experience of entrepreneurial cities, as well as the widespread practice of strategic city and regional planning there has also informed the CDS policy (Cities Alliance 2002). Urban managers in poorer cities are being drawn into circuits of consultancy and mutual learning that frame urban government across the world, circuits that are deeply marked by technical and more directly political forms of power.

It is the politics of city-development strategies, though, that are central to understanding their significance and potential. Early evaluations of CDS processes reveal that to be effective and to achieve the stated policy imperatives to scale up urban development and to create a replicable instrument of international urban-development policy requires a stable, consistent political context (GHK Consultants 2002). As Stren (2001: 196) notes, 'the process involves technical expertise (particularly in the early stages) but is essentially a political exercise, one of the most essential ingredients of which involves participation by major stakeholders in the elaboration of a "vision" for the city.'

The city is being drawn into view, then, through the process of building a strategic vision across the range of political interests, social needs and economic activities that make individual cities distinctive. But while the CDS process territorialises urban politics at a city-wide scale, the fragile nature of coalition politics in cities is quickly apparent in actual CDS experiences, as are the many obstacles to the representation of the interests of poor communities at a metropolitan level (GHK Consultants 2002, DfID 2001). This undermines, as Stren suggests, more technical approaches to formulating city visions that anticipate easy consensus-building and rational planning on a city-wide scale. The example of a CDS in Johannesburg, which we will explore next, illustrates the deep divisions that can characterise urban politics and the difficulties of enabling political representation at the city scale. Building towards an inclusive, city-wide development strategy is challenging, partly because of often stark inequalities and partly because urban politics is conflict-ridden: divisions amongst different interest groups within cities are often deep and intractable. Working with and across these tensions is the stuff of urban politics and represents the challenge facing all cities that embark on any participatory process of city-visioning.

The CDS initiative envisages an opening of urban policy to the diversity of interests and activities in distinctive cities and imagines wide participation towards building a consensus vision for the future development of a city.

However, achieving this goal is likely to be very challenging in the face of uneven organisational capacities amongst participating groups, local political conflicts and electoral uncertainty (GHK Consultants 2002). Furthermore, as we have seen, CDSs are also shaped by the power and interests of international development actors. And as Stren hints, the city that comes into view through a CDS process is strongly influenced by the intellectual and policy discourses that frame technical exercises in city mapping, usually prepared to inform the stakeholder deliberations (Parnell 2002, GHK Consultants 2002). This is likely to limit the range of options available to participants in CDS forums by framing the terms of the debate in advance.

However, the following discussion of a CDS in Johannesburg illustrates that despite the substantial political problems associated with drawing together diverse interests to negotiate a vision for the future development of a city, the exercise itself can nonetheless initiate new ways of seeing a city. To some extent, then, the CDS can bring the ordinary city into view. In Johannesburg, the process of formulating a CDS meant, for example, that neither developmentalist analyses nor global-city ambitions were able to be easily deployed. Both played an important role in framing and limiting the potential for imagining Johannesburg's futures (Robinson 2003a). But the very act of taking a city-wide view meant that these partial and limited accounts of cities and their economies were quickly redundant. Bringing the city back in through CDSs, although, as we will see, a politically complicated and even compromised process, has the potential to open up a space for new, imaginative accounts of ordinary cities such as Johannesburg.

JOBURG 2030: THE POLITICS OF CITY VISIONING

They [the World Bank] were debating what a city development strategy is and, and we gave them a case study, pretty much at the same time. I presented what we were doing in Johannesburg at [. . .] their Urban Forum, whenever it was, a couple of years' ago. And they said, that's a City Development Strategy, and I said, yes? It is? Oh, okay! [*Laughs*]. But they, they had the sort of basic framework in mind, and essentially it's about trying to look at a city comprehensively. So you look at the economy, the physical infrastructure, the people, the environment, the culture, security, legal framework or legislative framework, regulatory issues, and you know, whatever else you want to look at, all at the same time. And 2010 was probably the closest attempt at trying to do that, er when we were doing it. Maybe other people had done it, you just didn't find any evidence of it. And we learned lots, lots, lots of lessons from trying to do it in Joburg. But I think if we were doing it again, and we should, it would be a much better process.[14]

At the beginning of the post-apartheid era, Johannesburg had experienced many years of underinvestment in infrastructure in poor areas. The unevenness of development across the city had been much exacerbated by a highly fragmented and racialised form of local government – during the apartheid era there were as many as ten administratively separate local authorities governing different parts of the city and responsible to a variety of different sections of provincial and national government (Beall et al. 2002: 69, see Figure 5.1, Map A).

A laissez-faire approach to growth in the wealthy and white northern areas of the city had revitalised the financial and business-services sectors of the urban economy, especially as Johannesburg emerged as a major regional centre for business activities throughout Africa during the 1990s (Beavon 1998, Tomlinson 1999). But this expansion vastly overstretched the available infrastructure there. At the same time, a new African National Congress local government faced the strong demands of their support base in the local electorate to redress the decades of poor service delivery and lack of housing provision for African people: as much as one fifth of the city's population was estimated to be living in informal settlements or backyard dwellings in formal townships, and more than 10 per cent of the city's population has to make do without access to most basic services on their site (see Table 5.2). The

Table 5.2 Some estimates of access to basic services in Johannesburg

Service and level of provision	1996 All races,[a] percentage	1996 African residents,[a] percentage	Current city-wide service provision at selected minimum service level[b]	Monitor Group (2001), percentage
Water: at least a tap on site	89	84	Water: communal standpipe	96.4
Electricity: at least connection on site	85	79	Electricity: connection	85
Housing: formal house	90	82	Adequate housing (as opposed to informal (116,000 households) or backyard (106,000 households))	67
Sanitation (at least flush toilet on site)	93	88	At least ventilated pit latrine	84

Notes:
[a] Beall et al. (2002: 154–6), based on Annual Household Survey.
[b] The decision about minimum service level (MSL) is a crucial one, and not only in terms of cost or estimates of the challenge to the city. Critics of the MSL of a VIP latrine, for example, point out that the environmental and other externalities of this choice would outweigh the savings over full water-borne sewerage provision to households (Bond 2003). Monitor estimated that meeting these MSL service targets could cost between 3.7 billion rand and 4.5 billion rand in initial capital outlay, not including ongoing subsidies for a free lifeline service level to the poorest households.

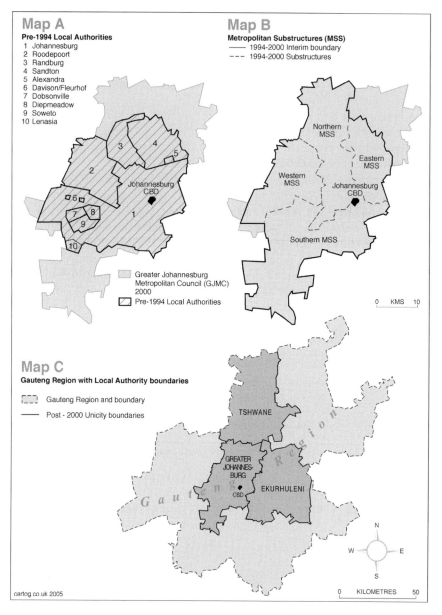

Figure 5.1 Maps of changing local government administrative areas in Johannesburg, 1990–2001. (Adapted from Beall et al. 2002 and City of Johannesburg 2001.)

competing demands of globalisation and developmentalism are seldom so starkly framed and, certainly, not often in such a politically charged fashion. In the post-apartheid context, where the city has been so brutally divided for so many years, the importance of taking a city-wide view of urban futures was undeniable.

The background to city visioning

It was in this context that the city managers in Johannesburg embarked on a variety of initiatives to envision the future of the city. The provision of infrastructure both within townships and also to support new connections between segregated townships and economic opportunities in the central city and other economic nodes was a priority for the new local authorities. From 1994, post-apartheid Johannesburg was amalgamated into four municipalities, with a relatively weak overarching metropolitan council, until 2000 when a unified metropolitan government was formed (see Figure 5.1, Map B).

With the first round of post-apartheid local authority investments to address the infrastructural backlog in poorer areas, though, it was quickly apparent that Africa's wealthiest city was in a perilous financial position.[15] Controversially, the city was effectively placed under administration in September 1997, and a caretaker body (a 'Committee of Ten'[16]) was established to see the city through a period of substantial institutional restructuring. This involved a massive reduction in capital expenditure, down from 1.6 billion rand to 300 million rand within six months.[17] Reviewing the finances of the city also opened the way for a major restructuring of the forms of service delivery with a stronger emphasis on public-private partnerships. This included agency relationships with entities managed by private companies but in which the council retained full ownership, corporatisation of activities within the council and the wholesale privatisation of about 5 per cent of the council's activities.

At the same time all appointments were reviewed along with the entire structure and procedures for council business. This restructuring process, which coincided with the introduction of a 'Unicity', a single metropolitan-wide administration, drew major public opposition from municipal unions and a loose coalition of organisations drawn together under the rubric of the 'Anti-Privatisation Forum' (see Ngwane 2003). As Parnell notes, in popular memory, the restructuring process, known as iGoli[18] 2002, 'is associated less with the creation of a metropolitan structure, than it is with the privatisation of services. So it is widely described as the neo-liberal edge of local government transformation in South Africa' (2002: 18).

For the city administrators, just seeing the controversial restructuring processes through was not enough. As the chair of the 'Committee of Ten' Kenny Fihla put it, by 1999 they were asking themselves:

> Yes, we are stabilising the financial situation, yes we are dealing with administrative issues, but to what purpose? In the short term, yes, we need to have a healthy financial position, we need to have a good and efficient administration and so on, and then what, what do you do once

all of this works? What will this efficient and good administration do [. . .] we have to start to think about Johannesburg beyond the crisis.[19]

A new city manager had been appointed in early 1999 by the Committee of Ten, now known as the Transformation Lekgotla, 'to lead the City into a new era of sustainable service delivery' (City of Johannesburg 2001: 42). Ketso Gordhan and his team of new senior officials initiated both the short-term restructuring of the city and a longer term visioning process, which was labelled 'iGoli 2010'.[20] It was this experience that the President of the World Bank, James D. Wolfensohn (2001) claimed had inspired the World Bank's major new initiative to promote urban development through CDSs. Here, a locally familiar form of city-wide representation, involving community, business, trade-union and council representation, was mobilised to initiate a 'Transformation Partnership'. Over 300 people attended a summit to present iGoli 2002 and to launch iGoli 2010 in August 1999. As Kenny Fihla noted at the meeting, once they had got the basics right there was 'a need for a longer term approach [. . .] to achieve the establishment of Johannesburg as Africa's world class city'.[21] And even at this early stage, basic needs and economic growth were identified as the two main concerns for long-term planning: the challenges of the diverse, ordinary city were already on the agenda. A steering committee was established with representation from across the different groupings and, after initial protests around iGoli 2002, broad support was secured for the initiative from all major constituencies.[22]

Johannesburg had a long history of building city-wide forums for negotiating across political differences both during the apartheid era and during the lengthy transition from apartheid[23] (City of Johannesburg 2001, Beall et al. 2002). In fact, all new local government dispensations around the country were negotiated through transitional councils on which broad political representation was legally mandated (Robinson 1998). If CDSs need strong organisational capacities across a wide range of constituencies to succeed, Johannesburg might be one place where this strategy for visioning city futures could be expected to work. These histories of negotiation had built strong capacities across constituencies to engage at a city-wide scale over crucial metropolitan planning and economic issues.

The politics of city-visioning brings things to a sticky halt

Three processes undermined the effectiveness of the political forum that was intended to negotiate and oversee Johannesburg's long-term visioning strategy, iGoli 2010. First, many capable local activists had moved into local, provincial and national levels of government after the first post-apartheid elections. Community movements and local unions had lost many of their best organisers and were relatively dormant. The Steering Committee appointed a technical-support team for community groups and, although participation from civic organisations themselves was practically nonexistent, the technical representative arranged focus groups with the support of local organisations and fed some sense of community priorities into the process, alongside a more quantitative 'customer survey'.[24]

Second, the bitter conflict between the city council and the municipal unions had spilled over into wider discontent over the iGoli 2002 process. Initially, the unions and community organisations refused to participate in iGoli 2010 unless the institutional restructuring issues were placed on the table again. This was sure to stall any negotiations over long-term visions. In the end, even though an agreement to continue discussing institutional restructuring issues at the iGoli 2010 meetings was reached behind the scenes, thereby clearing the way for union support of the process,[25] the unions withdrew anyway. Political disapproval also compounded difficulties of organisational capacity amongst community groups and ensured that there was very limited community participation in the stakeholder committee, apart from a small number of mostly white ratepayers' organisations.[26]

Finally, the external consultancy team (the Monitor Group[27]) that was gathering the data required to inform visioning discussions was delayed in submitting the final report, and a new round of local government elections at the end of 2000 saw a new mayor and the creation of a unified local authority for the whole city. In the end, the political component of this ambitious and very well-resourced visioning process was undermined by diminished capacity amongst community organisations, popular protest at institutional reform and the timing of the electoral cycle.

Participants and drivers of the process were disappointed. One of the ANC councillors involved in the 2010 Steering Committee suggested that he 'would die for it to happen' as planned, since they had been talking about 'getting people involved in determining their own future' for years.[28] Participants in the Steering Committee were frustrated that after all their hard work and enjoyment of the process, there had seemed to be no output.[29] Even some members of the oppositional South African Municipal Workers' Union (SAMWU) appreciated in the abstract the potential value of engaging the council over the long term planning of the city.[30] But the conflicted institutional and political situation meant this was not to be.

But the visioning process continues . . . as does the politics

However, this didn't prevent the generation of strong visions for the future of the city of Johannesburg. The iGoli 2010 initiative was transformed into an internally driven process of vision formation within the city council that resulted in the production of a widely circulated and influential policy document, *Joburg 2030* (GJMC 2002). Extensive consultations were conducted amongst senior officials and elected leaders of the council to identify a strategic focus for the city, drawing on the consultants' report that had been prepared by Monitor for the 2010 process.[31] This highlights the potentially important role that the formal institutions of local democratic government can play in articulating something of the diversity of interests across the city.[32] The institutional politics of 'actually existing democracy' – as opposed to issue-based participatory forums – functioned here to negotiate across conflicting agendas and interests within the city. In the end, and despite the powerful effects of the various consultants' discourses and analyses, the need for local government to take a city-wide perspective, even without a dedicated participatory process, did enable the production of a distinctive vision for the future of the diverse, complex and divided city of Johannesburg.

The consultants' analysis and assessment of Johannesburg's economic potential made an important contribution to setting the terms of the debate about the future of the city. Monitor, for example, pointed to the existence of a gap, between where Johannesburg is now and where it would like to be. To overcome this gap, they suggested, stronger economic growth was the only solution. In their presentation to the newly elected council in December 2000, the following vision for Johannesburg was distilled from the process: 'becoming a *world-class city* defined by increased *prosperity* and *quality of life* through sustained *economic growth* for all of its citizens' (Monitor 2000 emphasis in original).

The driving force of the economy was emphasised and certainly gave strong encouragement to politicians and officials eager to make economic growth a priority. However, the diversity of the ordinary city made this difficult to achieve without attention to the wider city context. The needs of the poorest residents were not only electorally important, national urban legislation mandated developmental local government and demanded strong attention to delivering basic needs. Thus the strategic vision for Johannesburg had to adopt a more nuanced perspective on how to secure a better future for the city. Both the consultants and politicians acknowledged that the city did not have the luxury of not delivering basic services: growth and service delivery had to be considered together. From the very beginning this had been completely obvious to the City Manager at the time; for him, growing income in the city and delivering services were both directed to the same political imperative: addressing poverty.[33] Monitor repackaged this basic insight to suggest that investment in service delivery would provide a broad platform for competitive growth. The economy, then, was portrayed as the underlying priority for the city, but any political support for economic growth had to find a way to include attention to basic service delivery. When this pro-growth agenda was even more vigorously pursued after the 2000 elections, political opposition from within council structures nonetheless ensured a more redistributive, service-oriented vision of the city's future. Senior councillors and officials associated with local administrative regions[34] and the Mayor, who was ultimately responsible for the electoral success of the ruling ANC party in local government, objected to the low profile of redistributive and social agendas in early drafts of the *Joburg 2030* document (Parnell 2002). Although efforts to solicit public input in this process were rather perfunctory and post hoc,[35] the representation of the interests of the poor through the electoral system meant that a city-wide perspective, incorporating both glob-alising economic growth and service delivery, was still achieved.

The politics of bringing the city back in

Bringing the city into view within CDSs is a profoundly complex and con-flictual political process. The Johannesburg example mirrors experiences recorded for other cities that have embarked on some form of CDS, although these have usually involved stronger donor or multilateral agency inter-ventions. The role of electoral and other political cycles, the difficulties of negotiating political conflicts, limited organisational capacities of com-munities, and the powerful role of consultants are common themes in accounts of the difficulties of CDS processes (Parnell 2002, GHK Con-sultants 2002). Furthermore, decision-making and urban management are taking place within a strongly neo-liberal policy environment and, as in Johannesburg, this has the potential not only to cause political conflict but also to encourage systems of financing, governance and service delivery that are not very redistributive. The outcomes of these processes are certainly highly politicised and historically contingent. But simply by bringing the city back into view, this initiative offers new opportunities for responding to the diversity and complexity of cities. It sets out the need for urban-development policies that address the diversity of interests and activities in ordinary cities.

CONCLUSION

This chapter has explored the convergence of enthusiasm for bringing the city back in to urban theory. This is illustrated by policy initiatives to support the formulation of CDSs, but it is also evident in emergent accounts of cities and city-regions more generally. The imperative is to develop understandings of urban processes that take seriously the diversity of needs and activities in a city. This can encourage a political and policy-relevant search for visions and strategies for city futures in which concerns with economic growth, sustainability and the needs of poor people can be considered together.

However, and as the detailed case study of Johannesburg demonstrates, the political dynamics at work in policy making at the city-wide scale are shot through with unequal power relations and have the potential for high levels of conflict. The conceptual and practical field of visioning the city's com-plexity and distinctiveness is filled with powerful agents, often limited by a restricted analytical focus and subject to prolific political contestation. These processes also take place within a strongly neo-liberal international policy environment. With such profound political influence on the production of city visions, the CDS process itself is clearly not going to necessarily deliver the best accounts of ordinary cities and their possible city futures. None-theless, there is much that urban theory can learn from and potentially con-tribute into this fraught field; indeed, I would argue that there is a strong imperative for scholars to contribute to these debates, if only to challenge some of the assumptions that underpin the technical advice offered during the process. In order to do this, though, the city, in all its complexity and diversity, and with all its antagonisms and contradictions, must be brought more fully into view within urban studies. Accounts of cities that stress only externally focused networks and flows, or that take extremely partial views of the city and neglect the territorialised field of the political or the integrating effects of claims for redistribution have little to offer here. By contrast, CDSs, for all the political contestation and neo-liberalising policy limitations that they entrain, indicate that it is the coexistence of globalising and poor areas of cities that ought to draw our attention. It is the ordinary city, not the global city or the Third World city, that needs to be the starting point for analysis and policy.

6 City futures

Urban policy for ordinary cities

INTRODUCTION

Ordinary cities – distinctive, diverse, contested – pose the question of urban development in new ways. Ambitions to improve life in ordinary cities require strategies to enhance a wide variety of urban environments; encouraging the expansion of urban economies will mean paying attention to the diversity of economic activities that cities bring together. The previous chapter explored how city-visioning processes can help to bring into focus both globalising economies and poor neighbourhoods, overcoming the division in urban studies between accounts of globalisation and development. This chapter builds on this by considering how interventions to shape city futures can respond to the distinctive features of each city context. In particular, it is the specificity of the local social and political context and the economic diversity of cities that can form the basis for imagining new urban futures.

The chapter, then, examines the consequences for urban policy of seeing all cities as ordinary. And it will make the case for the benefits – political and intellectual – of learning from the experiences of cities often kept apart by the divisive categories of urban studies. Thinking across a diversity of urban contexts can help to ensure that academic analysis and policy advice don't fixate on a few exceptional cases. Building on Scott and Storper's (2003) suggestion that urban development policies in poorer countries can learn from regional economic analysis in wealthier contexts, I argue here that it is also very important for wealthier cities to be open to learning from the experiences of poorer cities. One of the ambitions of this chapter, then, is to demonstrate the possibility of a more cosmopolitan form of theorising and policy-making which, like the inventive and innovative urbanisms explored in Chapter 3, draws inspiration from a range of different contexts.

I begin by suggesting that urban policy-making within a post-colonial frame depends on an account of modernity in which all cities are thought of as potentially creative and dynamic. In order to envision a distinctive future for a city such as Johannesburg, city managers and residents need to confidently assume the modernity of their city. Without a strong sense of the city's potential dynamism and creativity, imaginations about urban futures are truncated, perhaps by consigning futures to the limited imagination of developmentalist interventions, or through a narrow focus on globalising sectors of the economy. Avoiding this will involve appreciating the distinctiveness of a city, and it will require an openness to the diverse transnational networks into which cities are inserted. Urban managers, then, confront the distinctive challenges of their city, even as they draw on wider circulations of knowledge about urban development and liberally borrow ideas from across the world. Inspired by a cosmopolitan account of modernity, then, we can insist that all cities have the potential to shape distinctive future trajectories, despite the unequal power relations that characterise the world of international urban policy.

Within this framing, any strategy aiming to shape the future of a city will need to confront its distinctive social and political context. The demands of the diverse residents of cities press on policy-makers, even as they find they also need to encourage economic growth. Since promoting the urban economy is increasingly a political responsibility of local governments across the world, and visions for the futures of cities generally try to address this, the second section of this chapter explores how the competing demands of economic growth and attending to the wider social and political context of the city are addressed in different contexts. This is a core policy challenge in a neo-liberalising policy environment, and there is much to learn across wealthier and poorer cities about how these agendas are reconciled (or not) in different ways in different cities.

It is not just the specificity of the social and political context of cities that inspires distinctive and sometimes creative urban interventions; each urban economy, too, is distinctive as a result of the array of different kinds of activities that come together in any particular city. The ordinary city brings into view the very wide range of economic activities and ways of making a living that make up city economies. The final section of the chapter makes the argument that, far from being a disadvantage for urban development, there is emerging evidence that economic growth and innovation are fostered by diverse economies. Specialisation and diversity both have the potential to encourage economic expansion, which is good news for many large poorer cities with a wide range of economic activities but also for some of the wealthiest cities in the world. There is the opportunity, then, to consider policy initiatives in which cutting-edge globalising sectors, older industries and businesses, as well as less dynamic but employment-rich activities are important. Also, fostering connections across formal activities and the informality of many city economies, and incorporating transurban networks of varying reach might be important tactics to support urban economic expansion. These strategies have the potential for urban development policies to support economic growth and, at the same time, to have redistributive effects in the wider city.

Thinking of all cities as ordinary, then, has important consequences for how we imagine economic growth to occur, and holds the potential for generating new strategies for promoting development in cities, strategies that appreciate the uses of diversity and that can respond to the distinctiveness of different cities.

URBAN POLICIES, INVENTED AND BORROWED

Strategic visions for the future of cities – the broadest and most long-term arena of urban policy-making – have been popularised in many different cities by international urban consultants and international development agencies alike. They also have a longer history within the physical planning of cities (see Healey et al. 1997). As with the visioning process conducted in Johannesburg, these exercises are often wide-ranging explorations, first, of the dynamics and dimensions of the city itself, and second, of potentially useful approaches to urban development that have been tried in other contexts. They highlight the predicament of urban policy-making in general, within a globalised policy environment: the importance of sensitivity to the specificity of the city being considered and the need to respond to that city's location within wider networks of circulation, competition and power.

But the differential location of cities within these wider power relations is crucially important. Especially where influential consultants or agencies with money to donate or loan can swerve urban agendas, there is a need for a robust appreciation of the potential vitality and dynamism of all cities, and of the benefits of mobilising and building on their distinctive capacities. Here I suggest it is most helpful to return to the discussion of urban modernity, developed through Chapters 1 to 3, which insisted that creativity and innovativeness, in so far as they are qualities associated with cities, are a product of circulations through and beyond cities and can be appropriated by all cities: they are not the property of wealthier cities. The foundations, then, of creative urban policy-making are to be found in a strong appreciation of the modernity of all cities, and an awareness that much of the dynamism and innovativeness of urban contexts stems from the prolific circulations of ideas, resources and people into cities and beyond them. The city-visioning processes in Johannesburg, which were introduced in Chapter 5, demonstrate this very clearly.

Urban modernity and city futures

Imagining a distinctive future for the city of Johannesburg, as we saw in Chapter 5, has at times been caught in an impasse between globalisation and development. These often mutually exclusive circulating discourses and practices could work to limit an appreciation of the specific complexity and diverse demands for intervention that come together in that place. Agendas for pro-poor development, commissioned by international development agencies, have arrived on the desks of senior officials at the same time as reports that stress the potential for Johannesburg to be a 'global city'.

Promoting competitive economic clusters and delivering a basic sewerage system across the city equally attract the attention of policy-makers here. But rather than being caught within the dichotomies of these externally driven agendas, as they turned their attention to the challenge of post-apartheid urban development, city managers had to hand a repertoire of cultural representations of the city which have been able to ground their imagination of distinctive futures for this place of gold, *iGoli*.

The slogan and logo for the iGoli 2010 Partnership, for example, captures the way in which a range of different agendas came together within a powerful appreciation of the agency of the city to inform a particular vision of what Johannesburg might be (see Figure 6.1). With lines of power shown radiating across the sub-continent, Johannesburg's strong role as a regional centre is highlighted. The city, then, is actively influencing developments far beyond its physical extent and is portrayed as a source of powerful agency on the continent. It is the framing role of Africa which, juxtaposed with the slogan, 'Building an African World Class City' indicates something rather different than simply a city that undertakes 'global-city functions' for the African region. The awareness of the city's African cultural heritage and, increasingly, its connections through migration to many different parts of the continent, sets a different gloss on the desire for status and excellence that the 'world class' ambitions signify. For some observers, the Africanness of Johannesburg speaks not only to the vitality of life there, but also to the difficulties of managing a city characterised by informality, often apparently disorderly public spaces and the sometimes overwhelming challenges of maintaining the infrastructure of the city. Something of the distinctiveness of the city is brought out with this strong African framing – but so is the search for a political and cultural legitimacy for the elite managing the visioning process.

In the logo design, incorporated in the word 'iGoli' itself is an icon of Johannesburg's modernity: the communications tower in the inner city area of Hillbrow. This tower is reused in the later *Joburg 2030* document (see Figure 6.2), and is ubiquitous in the city council's publicity – web site, business cards, the covers of routine reports. While the other elements of the iGoli 2010 image speak to the diversity and specificity of the city, the tower stands out as a sign of the city's potential, in the same way perhaps as the Petronas Towers in Kuala Lumpur appropriated the verticality of a certain version of urban modernity to signify its place on the global stage. Through this image, post-apartheid Johannesburg's city managers are articulating an ambition to create a modern, world-class city today – and are signifying the already existing modernity of that city. In fact, post-apartheid urban managers share both this ambition and this icon with the apartheid engineers who, forty years before, had built this and another towering structure on either side of the inner city.

In January 1962, a new FM Tower was formally launched as the site for radio broadcasting using the new 'frequency modulation' system in Brixton

Figure 6.1 Logo of the iGoli 2010 Partnership. (Source: GJMC Pamphlet, *Gateway to a Better Life*, 2001.)

on the Western edge of central Johannesburg. The whole nation could now receive broadcasts: a new era in mass public broadcasting was initiated and African-language broadcasting expanded substantially with the founding of Radio Bantu, used to full advantage by apartheid propagandists. The tower was apparently an astonishing feat of engineering and construction and was operational only two years after it had first been proposed. It was opened with fanfare in September and named after the Minister of Posts and Tele-communications at the time, Dr Albert Hertzog, and its appearance on

Figure 6.2 The Hillbrow Tower on the cover of *Joburg 2030*. (Source: GJMC 2002.)

Johannesburg's skyline was much celebrated. Visitors were whisked to viewing platforms at 585 feet (still 200 feet away from the top of what was then the eleventh tallest structure in the world at 772 feet (235m) in under a minute. There the city appeared stretched out before the visitor: 'breathtaking in its beauty and panoramic in its extent' (Telford 1969: 81). The even-taller Hillbrow tower at 883 feet (269 m), by contrast, dominated the eastern edge of the skyline of downtown Johannesburg. This sister to the Brixton tower was completed in 1971, originally named for J. G. Strijdom, another key architect of apartheid and welcomed visitors to a distinguished restaurant to take in the views.

For a while, the Brixton tower was the tallest structure in Africa, but this fact was trumpeted as an achievement of white rule; high modernism and high apartheid came together, then, in the form and meaning of these towers. The reappearance of the Hillbrow tower on report covers, business cards, web sites and magazines in the first decade of the twenty-first century is nothing if not curious, then. In some ways, these apartheid towers in their dated modernity can stage a 'dialectics at a standstill' for us, bringing the dynamics of contemporary urban policy crashing together with the histories of apartheid. They might challenge us to ask, for example, who is being excluded in these new imaginative configurations of Johannesburg's futures? Are racialised pasts perhaps being reinstated in an era that is ostensibly committed to a non-racial polity? But in many ways, the appropriation of these towers from the architects of apartheid to symbolise the future of an inclusive and dynamic city speaks more of the potential for icons of modernity to be appropriated at will and to carry ambitions of global success across the historical divide of apartheid and post-apartheid government.

Looking at the image of the Strijdom tower in Hillbrow as we prepare to open the local council's report and think of the future of Johannesburg, we might notice how it reaches far above the impressive neighbouring skyscrapers of the central city (Fig. 6.2). The glamour of the high modernist urbanism it embodies is being drawn on to convince new post-apartheid South Africans that we can be successful. Like the apartheid engineers of the 1960s, we want to believe we can build cities (and towers) as tall and as beautiful as any in the world. We are encouraged to buy into the competitive world of global capitalism with great enthusiasm, inspired by this apartheid tower. This iconic image of high apartheid is being used to put Johannesburg on the map of international trading and tourism and to stake a claim for a place in global cultural imaginations of modernity. It reflects a desire to be seen as part of the best of circulating modernity's achievements and to attract the economic growth that is understood to follow on from this global visibility.

Of course, the consequences of competing on the global stage of capitalism are by no means always to be celebrated. But the ability of policy-makers in Johannesburg to reach beyond any stereotyped view of the future of their city depended on generating a vision of their right to be modern, to occupy a visible place on the stage of the world of cities. The institutional energy, determination and capacity to generate a distinctive vision of Johannesburg's future found support, then, in the ability to imagine the city as modern.

Cosmopolitan inspirations

Drawing on the cultural energy of imagining cities as modern, and inspired by the complexity of ordinary cities and the array of political processes at work in them, then, distinctive approaches to urban development do emerge, but they are also shaped by wider circulations, ideas and influences. As one of Johannesburg's political leaders explained, they:

> developed a model on the basis of our own experiences, but yes, we also drew from the experiences of other countries. We explored Porto Alegre in Brazil for example. [. . .] We looked at Canada, how they moved into one city from their six counties. [. . .] there was a lot of work on the various governance models even before the creation of the metros. [. . .] There was the work done by Province and National Government and through academic institutions within the city, so we essentially tapped into virtually, every single source that we thought could be helpful to this process. [. . .] we were looking around, we were looking around. [. . .] we used our partnership agreement with Birmingham, to draw on the experiences in the UK and so on, and we sent people to visit Birmingham and London [. . .] we sent people to Australia to view their different models and options [. . .] so we kind of followed every single line where an example of what they've got could be useful for our own purposes and then adapt it so that it's useful.[1]

A strongly cosmopolitan imagination of possible urban futures was clearly at work in this energetic search for policy options in Johannesburg. However the ideas about urban futures circulated, whether by books, word of mouth, conferences, commissioned research, study tours, the advice of development agencies or consultants, the field in which Johannesburg's city strategies were formulated was intensely international. Insights and benchmarks from Singapore to São Paolo framed the search for a new vocabulary and insti-tutional practice of urban management in Johannesburg. Many of these related to the installation of proto-privatised institutional procedures for service delivery with a slim central managing body (influenced by the trips to New Zealand and the technical input of the World Bank). Popular ideas about what the city might become were also inspired by the idea of being an African city (Tomlinson 1999, Bremner 2000), although the ambition of the formal long-term strategic policy was that by 2030 Johannesburg would be more like San Francisco than Nairobi (GJMC 2002). Publicly, Johannesburg was setting out to be a 'World Class African City' and, as we have seen, that very slogan captured something of the complexity and diversity of the trajectories that would make up its future.

In coming to grips with the potential futures of the city, policy-makers operated in a whirlwind of ideas and influences. But, as noted in Chapter 5, it was the challenges of the ordinary city of Johannesburg – the complex, diverse and divided arena in which the municipality had to operate – that framed how these different influences came together. In developing a strategic vision for the city, the dual agendas of promoting growth and ensuring service delivery had to merge. The provenance of these agendas were quite different – the one more inspired by national neo-liberalising agendas and US urban economics and growth ambitions, the other by both local politics and wider developmental expectations. But, coming together in this city, they both had to be addressed. And, more than this, in planning for interventions in the economy of Johannesburg, the diversity of this economy – its stretching from micro- and informal enterprises to some of the largest conglomerates on the continent, from ageing manufacturing to high-tech software design – pressed upon policy-makers. What it might mean to develop the 'real' urban economy (following Harris 1992) in an ordinary city such as Johannesburg required at the very least a blending of different circulating policy agendas into something distinctive and particular to this unique city. We will return to these issues in more detail in the final section of this chapter in order to illustrate some of the ways in which urban policy might be refigured for ordinary cities. First, though, we turn to consider in a more general sense how the distinctiveness and diversity of ordinary cities might influence understandings of urban policy.

CONTEXTS OF GROWTH: SERVICES, POVERTY, COHESION

> What is controversial is the direction of local government activity: whether it should be directed solely at efficiency, reinforcing the current distributions of wealth and power, or whether it should play a redistributive role, creating a minimum standard for quality of life for all its residents.
>
> (UNCHS 2001: 39)

As with Johannesburg's policy-makers, urban managers and scholars elsewhere, concerned to understand and promote the urban economy, have had to pay close attention to the wider context within which economic growth takes place. This has taken different forms in different cities, especially since the balance between growth agendas and wider urban concerns is a deeply political issue. Local governance structures and the relative capacities of various constituencies to organise around these issues are key determinants of the different strategies adopted by different cities for promoting economic growth while maintaining social cohesion, addressing poverty or delivering the basic services required for safer urban living. The political and policy challenge of developing diverse, ordinary cities has kept sharp debates about these issues on the urban agenda for many years, even in those situations, such as in the USA, where there have been relatively strong imbalances between powerful capitalist interests and local community mobilisation.[2]

Institutional forms and political relations are crucial in explaining local political choices between emphasising economic growth and the demands of the wider urban context, but so too are the ideas about urban policy that circulate between cities. However, whether circulating discourses about urban policy are taken up in different contexts or new urban development ortho-doxies invented in particular cities is also partly determined by the political and institutional context (as we saw in the case of Johannesburg). The inven-tion of participatory budgeting in Porto Alegre, for example, was a fortuitous outcome of a strong neighbourhood organisation that pressed an at-first-hesitant ruling party towards greater participation and a weak political opposition that failed to notice the initial successes at this level (Abers 1998, Goldfrank 2003). The Port Alegre model has circulated widely as 'best prac-tice' in urban governance although often the local political context elsewhere has militated against its effective implementation, even in other Brazilian cities (Acioly 2001). Conversely, in some cities weaker local government capacity could mean that national governments, powerful donors or external interests are more able to shape agendas, although local agency can still shape both outcomes and the pace of implementation (see, for example, Myers forthcoming). But in many cities around the world, including the largest and most wealthy, fragmented and dispersed forms of government can make formulating and implementing city-wide agendas difficult and can hamper local government's ability to intervene in shaping either economic or social agendas.[3]

The dynamics and difficulties of urban governance set the context for urban development policies in different cities, shaping the relative impor-tance of growth and wider urban agendas. There is scope, then, to think across the experiences of different cities in terms of the political dynamics and discourses that have determined their trajectories. Rather than assuming that different kinds of cities occupy incommensurable policy and political realms, a more comparative or cosmopolitan form of urban analysis might enhance the possibility of engaging critically with urban growth agendas and imagining opportunities for tying urban interventions for economic growth more closely to a politics of redistribution.

This comparative exercise has already been undertaken in relation to cities in the USA and the UK.[4] Cities in the USA have a long history of boosterist initiatives to stimulate local economic growth, facilitated by the local dependence of certain kinds of economic interest (utilities, banks, property developers, locally embedded manufacturers), although also motivated by electoral goals to deliver on goods and services to local people, including protecting service levels through suburbanisation and secession.[5] But it was only in the 1980s, with the liberalisation of government in the UK and the EU that more entrepreneurial city government began to emerge in these contexts. Despite drawing on the experiences and practices of US cities, the challenge of continuing to address questions of social equality, redistribution and poverty or social exclusion at the same time as striving to make cities more competitive, has drawn considerably more attention in the European context. A substantial literature has emerged and extensive funded research has been undertaken on the relationship between competitiveness and cohesion in British cities. Thus, a specific attempt has been made to engage with the generalised concern with balancing growth and the wider urban context, here within the distinctive welfare-state-oriented political order and strongly influenced by the circulation of ideas about poverty as social exclusion.[6]

In the UK, the strong centralisation of redistributive activities and of funding for social needs has meant that the agendas of growth and redistribution are often effectively disarticulated at the local level. Combined with an often fragmented local government structure,[7] the relationship between strategies to secure more economically competitive urban centres and to ensure the continuation of the relative social cohesion bought by the welfare state has been rather tenuous. None of the extensive research shows any clear or direct links between initiatives to promote competition and the existence or decline of social cohesion. Certainly the welfare state cushions even the least successful cities from the serious social consequences of economic decline (Turok et al. 2004). Some writers postulate that there might be certain 'thresholds' beyond which social problems start to affect economic activity – as crime rates rise, for example, or long-term unemployment affects more of the population (Buck et al. 2002). But in the absence of any comparative research, the UK studies have no ideas as to what these might be.[8] They conclude that social cohesion is a 'good' in its own right and should be supported and promoted for itself, rather than for the purposes of enhancing competitiveness (Buck et al. 2002, Turok et al. 2004).

The balance of local support for economic growth or social well-being seems to be affected by a range of factors, including the political party leading the council and the relative success of local networking with key actors (Harding et al. 2004). Leverage of external funds from national government or from the EU (which requires high levels of organisation and time) can determine development paths. But the basic incentive structure for local authorities strongly favours personal council tax as an income source (making higher-income residents desirable to councils). Furthermore, local business earnings are redistributed across the country (making employment generation a less attractive alternative). This has contributed to a high level of urban regeneration projects (as opposed to employment-generation initiatives) and has led to charges of 'cappuccino urbanism', especially in the highly fragmented London region, perhaps reinforced by a strong design influence in some national urban policies (Buck et al. 2002, Turok et al. 2004).

In very poor cities, the demands for intervention in the wider urban context are, if anything, stronger while the resources are far more meagre. Varying levels of local taxation, as well as the range of services provided at the local

level, in addition to the variability of national and international contributions to urban development make any comparison of local income
and expenditure across national contexts very difficult, although Table 6.1
presents some statistics that have been compiled by the UN from the early
1990s.

Despite these vast differences in resource, the dilemmas and politics of
promoting economic growth and addressing liveability issues in the poorest
cities around the world resonate with the demands of governing some of the
wealthiest. Lusaka City Council, for example, in its 1999 mission statement,
sought to secure a strategy that could 'improve the quality of life for all those
who live, work, visit or conduct business in Lusaka'. Part of the strategy
involved trying to secure a stronger revenue base for the city and attempting
to support economic activity in the city in order to be able to deliver better
services. Planned services included orphanages and attempts to control
communicable diseases – of heightened significance in a country with many
thousands of AIDS orphans. But the imperative to secure better services and
the need to ensure some support for economic growth in order to make this
happen was also considered important in this context (Lusaka City Council
1999)[9].

In contrast to wealthier city contexts, developmentalist initiatives to
address poverty alleviation in some of the poorest cities have been alert to the
close intertwining of social welfare and economic activity. It is no surprise
that for many poor people access to jobs and employment of some kind are a
priority – as a community survey held as part of the CDS in Johannesburg,
for example, illustrated. More generally Hall and Pfeiffer note, that 'Social
development and economic development, especially in the informal sector,
overlap to a considerable extent: lack of resources or of income is the
dominant cause of social problems' (2000: 237). More than this, in drawing
on a 'livelihoods' approach to address urban poverty, development scholars
and policy-makers have framed a field in which building the resources,
capacities and 'asset base' of the poorest urban residents forms the focus of

Table 6.1 Local government revenue and capital expenditure per annum by region
(1993 data). (Source: UNCHS 2000.)

Region	Revenue per person, US dollars	Capital expenditure per person, US dollars
Africa	15	10
Arab States	1,682	32
Asia	249	234
Industrialized	2,763	1,133
Latin American Countries	252	100
Transitional	237	77
All cities	649	245

poverty alleviation strategies (Moser 1998, Rakodi with Lloyd-Jones 2002). In these interventions, assets for livelihoods can include houses, skills, labour as well as social networks, family relations and community-based social capital – although, as Moser and others have argued, social capital does not always work in positive ways. In the poorest cities, then, the entwining of social and economic issues is hardly in question, and appreciating these relationships is seen as a crucial foundation for framing supportive development interventions.

This relationship is widely accepted in the development literature (see Rakodi with Lloyd-Jones 2002), and local governments and development organisations are eager to find effective ways of expanding local economic opportunities through supporting the livelihood strategies of the poorest city dwellers or, at the very least, not disrupting existing activities. For cities more generally, the value of appreciating the close entwining of social life and economic activities, broadly defined, is strongly reinforced by learning from the experiences of poorer cities. As a result, and as Chapter 5 demonstrated, broader international policy initiatives such as the CDSs, aimed at cities in poorer countries, have tied the delivery of social and economic aims closely together – even within a neo-liberal policy frame.

In other contexts, the relative disarticulation of economic growth agendas and wider social needs has been the subject of plenty of contestation. In the USA, the impact of neo-liberalising policies since the 1980s has profoundly reshaped, or 'destructured' long-standing post-war social compromises in urban governance. Reframing national-, federal- and local-level responsibilities led to a 'new localism [. . .] in which states and municipalities began to adopt entrepreneurial strategies in order to attract external capital investments to their territories (with) devastating social costs' (Brenner 2002: 8). More recently, though, the consequences of these policies – growing inequality and segregation, a substantial spatial mismatch between public resources and social need as a result of metropolitan fragmentation, and the search for new spatial arrangements to ensure continuing economic growth – have all contributed to a rise in initiatives to forge metropolitan-wide institutions for planning, redistribution and stronger economic integration across the city. The politics of city futures in the USA, then, might be strongly shaped by contestations over the balance between economic growth and the wider city context, as both the possibility and the terms of these potentially reintegrating initiatives are determined by political struggle and also by institutional dynamics and place-specific social relations (Keil 2000, Brenner 2002, Jonas and Ward 2002).

The wider urban context is an important element in framing economic growth agendas for all cities because of the contested political environment in which interventions for growth must take place: the city is a social and political arena in which strategies for growth are necessarily embedded. So the political and institutional embeddedness of local economic agendas ensures that the wider context for growth can seldom be ignored – or not lightly ignored. Growth paths that neglect social needs and redistributional agendas can come unstuck, as crises emerge, or as reversals of globalisation, for example, expose serious social limitations (Firman 1999). Achieving live-able cities and social redistribution can also, as scholars writing within the context of robust welfare states remind us, be considered a good in itself, something that local (and higher-level) authorities ought to aim to deliver irrespective of its consequences for economic growth.

For many different cities, then, strategies for supporting economic growth and the wider city context are closely entwined. Studies in wealthier contexts are a salutary reminder that state interventions to support welfare and well-being are important in and of themselves, even in the poorest cities, while developmentalist studies from poorer cities suggest that other cities might benefit from paying much closer attention to the interconnections between the economy and social networks, urban resources and personal assets. In facilitating both social and economic activities through its general infra-structure, as well as through the sociability or alienations of city life, the resources of the city itself – the externalities of the collective life of the city – are a crucial component in enabling both economic growth and meeting wider social needs. These infrastructures and social resources are also at stake in more properly economic debates about the importance of agglomeration economies to the growth and dynamism of urban economies. The following section considers how the diversity of ordinary cities might speak to strategies for supporting urban economies, and in ways that are synergistic with a commitment to the wider social agendas that maintain urban livelihoods and wellbeing and are politically so important.

THE USES OF DIVERSITY: CLUSTERS, NETWORKS AND AGGLOMERATION ECONOMIES

The positive externality effects of the co-location of firms in cities, either in spatial concentrations of specialised sectors or more generally across the city, has been a core feature of urban economic theory. The infrastructures and physical fabric of the city and the positive benefits of co-location or agglomeration within cities has been understood to be an important con-tributor to the economic dynamism of cities. Marshallian economic districts, characterised by the broad externalities of collectively provided urban infra-structures, skilled labour pools, access to larger markets and the potential for forward and backward linkages in the production process, have been supplemented more recently by an appreciation of the opportunities for 'knowledge spillovers'. These include collectively produced opportunities for sharing innovative practices, either through increased interaction amongst firms or through circulating employees, thus generating innovations through inputs that are external to the firm. These 'soft' or 'untraded' interdependen-cies amongst firms have been thought to be very dependent on co-location (Amin and Thrift 1992, Gordon and McCann 2000).

The advantages of this spatial concentration of specialised sectors of dynamic economic activity for creating learning environments and opportunities for technological innovation has attracted substantial academic and policy attention in recent years.[10] Not only is there the question of explaining how it is that some localities and sectors become sites of creativity and economic dynamism, there is also the hope that these processes can be supported and indeed harnessed to promote local economic growth. Fostering successful clusters has become something of an orthodoxy of international urban policy and of economic development policy (Markusen 1996a, Schmitz and Nadvi 1999). Observations about the nature of a few dynamic, locally embedded clusters in specific contexts – primarily wealthy cities – have been translated into more general policy advice and framed substantial research endeavours (Malmberg 2003). However, although in its policy guise this initiative hopes to advise cities everywhere on a recipe for economic success, it could be argued that it has very limited direct applicability in many cities.

The primary interest in cluster-based policy advice and research has been motivated by a desire to identify ways in which wealthy countries can protect their relative economic advantage in a time of globalising production, by drawing on 'knowledge and quality that are not readily available in the less-developed economies' (Simmie 2004: 175). As Michael E. Porter and his collaborators note in relation to the USA: 'In healthy regions, competitiveness and innovation are concentrated in clusters, or interrelated industries, in which the region specializes. The nation's ability to produce high-value products and services that support high-wage jobs depends on the creation and strengthening of these regional hubs of competitiveness and innovation' (Porter et al. 2002: ix).

In these contexts, policy initiatives have been directed towards industrial systems and forms of social organisation that enable sustained technological innovation, in order to constantly benefit from the advantages of improvements in quality and production. As Michael Storper notes, though, there are relatively few regions where such innovation occurs. Nonetheless, he suggests that 'their economic impacts on the respective national economies are likely to be large' (1998: 91). This hopefulness is widely held, but the consequences of these development strategies for poorer, liberalising economies are not positive. Such competitive behaviour on the part of firms in wealthier economies has led to industries elsewhere being caught in a 'sandwich' between highly volatile price-based competition and the quality and brand dominance of firms in wealthier contexts, with great difficulties experienced in finding an export niche for local industries (Storper 1995, Altenberg and Meyer-Stamer 1999: 1700).

Although economic clusters have certainly been identified in poorer cities and regions, including very low-wage and informal contexts (Schmitz 2000, McCormick 1999, Altenberg and Meyer-Stamer 1999, Cawthorne 1995), the ability of wealthy cities to gain competitive advantage through high-growth, high-wage, innovative environments has considerably less purchase in these poorer cities. The dynamics of cluster development in poorer contexts are not always positive, for example, so the focus on 'technology' or 'industrial districts' as a basis for enhancing competition and industrial development may not always be very helpful. And, especially given the external dependence of so much industrial development in poorer contexts, the focus on locally based learning networks of collaboration and trust might also have very little purchase on understanding the spatialities of many ordinary city economies; these approaches may be misguided as to where in cities (or beyond them) creativity might be thought to be located. These limitations of industrial-district analysis, although highlighted by the experiences of poorer economies, are also important for developing a better understanding of the economies of all cities.

The 'Italian' model of economic clusters focuses on locally embedded networks where dense inter-firm relations based on trust, frequent interactions and durable relationships enable innovations and the competitive performances of these firms compared to those in other places. But this offers a very gloomy prognosis for the economies of poorer countries and cities.[11] As Michael Storper observed:

> The object of policy in the learning economy must not be simply to install hardware in a place and the skills required to operate it, but to set a nation or region on a learning-based technological trajectory in particular technological-economic spaces (ensembles of activities characterized by direct and indirect linkages). The task is a complex one, designed to keep the region moving from one point in a trajectory to another. As other regions in the world economy catch up and become capable of imitation, it moves onto activities which reflect scarce learning. This enables it to enjoy the economic quasi-rents of scarcity. The region, in other words, travels a technological trajectory so that it is a moving target, not a static one.
>
> (1995: 408)

As he concludes, this is a far more challenging requirement than 'the goal of import-substitution and national Keynesianism' (1995: 421). The need for strong institutional support in situations often characterised by relatively weak forms of governance (Schmitz 2000), the dependence on the presence of collaborative social relationships, high-end skills requirements and the enormous challenges of breaking into industries with strongly developed global competitors leave Storper, in this article, relatively pessimistic about the opportunities for even middle-income countries to engage in the innovative and creative strategies for learning that are associated with the most dynamic economic clusters. As he also points out, the search for 'propulsive' activities that, in their rapid and sustained growth, might offer opportunities to expand economies more rapidly and that, in their relative technological sophistication, might tie poorer countries into the circuits of more dynamic 'advanced' economic sectors is only one amongst many possible economic strategies for middle-income and poorer economies. But it is a strategy that is strongly prioritised in much policy advice to the detriment certainly of poor cities with limited resources where, as in the Johannesburg case, this involves a strong trade-off between future economic expansion and current gains in social well-being. This advice is also poorly applicable to many wealthier city contexts. As Simmie (2004) notes, only a minority of city-regions have the potential to generate the conditions required for such strongly localised learning and knowledge-related clustering.

If the policy advice of cluster-based analysis has limited relevance in many cities, what alternatives are there for policy interventions and analyses that might support cities – governments, citizens, firms – in their efforts to expand economic activity and enhance well-being? The broader academic critique of clusters within wealthier city contexts offers some signposts as does the example of Johannesburg that we have been exploring. Both point to the value of an emphasis on some different spatialities of innovation and economic expansion, including the importance of connections beyond localised clusters and the city-wide resources available to all firms, including the advantages of economic diversity.

There is an emerging consensus in the literature exploring mostly wealthy cities that suggests that the strongly localised cluster of small, flexible, inter-locking firms is an unusual phenomenon (Gordon and McCann 2000, Malmberg 2003). Forms of localised economies vary – as Markusen (1996b) explores in her discussion of alternatives to the 'Marshallian industrial district'. 'Hub-and-spoke districts' (dominated locally by one major firm, with substantial inputs and outputs to external firms as well as smaller local firms) and 'satellite platform districts', with externally owned branch plants dominating a diverse local economy, are also common forms of industrial organisation. In all cases, the opportunity for firms of all kinds to benefit from the broadest range of shared agglomeration economies is present. These might be most difficult to exploit in situations where externally focused plants might have fewer incentives to connect with the local context although, some analysts suggest that in very weak industrial situations externally owned firms might be an important, though sometimes limited, source of local learning (Schmitz 2000).

That there are a diversity of relationships both amongst local firms and stretching out to suppliers, buyers and external owners hints also at the limits to focusing only on the role of spatial proximity in innovation and economic dynamism (see also Raco 1999). And here there is hope that even cities and firms in relatively marginal locations in the world economy might find ways of forging innovative collaborations, building what Amin (2003) calls, 'relational proximities', through, for example, communities of practice drawn together not by co-location in the same town or area but by networks that combine occasional face-to-face meetings with intensive and ongoing communications (Malmberg 2003). As Markusen argues, 'Improving cooperative relationships and building networks that reach outside of the region may prove more productive for some localities than concentrating on indigenous firms' (1996b: 310). There would be no reason to assume that local ties are any stronger than interaction at a distance (Amin 2003: 124), and the 'relevant geography would not be local embeddedness (but) [. . .] relations of copresence [in which] people are able to internalize shared understandings' (Allen 2002: 463).

In addition to the significance of wider, 'global' networking, researchers have been drawn to explore the possibilities that the agglomeration economies of large cities with diverse economies offer to firms, outside of any specialised localisation economies or spatial clusters. There is renewed interest in this older argument and growing evidence that wider agglomeration economies – shared access to infrastructure, skilled labour, etc. – play a very important role in fostering economic activity of all types (Schmitz and Nadvi 1999: 1504, Duranton and Puga 2001). There has been a revival of interest in the suggestion – traced to Jane Jacobs (1972) – that in many circumstances it is economic diversity rather than specialisation that fosters inventiveness and creativity. The consequences of this are very important in all cities, but this is perhaps especially so in cities where the dual pressures of encouraging economic expansion and promoting redistribution must be faced with relatively few resources.

In London, evidence from firm surveys suggests that innovative firms there depend not on 'the tight local linkages found in the more regional production complexes of Lombardy or Baden-Württemberg, but access on a more "pick and mix" basis to the array of possibilities for connections available across the metropolitan region' (Buck et al. 2002: 119). The city is characterised by a rich diversity of different kinds of business environments and offers access across a broad city-region to a very wide range of services and potential collaborators. It also offers extensive access to international networks with easy opportunities for interaction (Simmie 2004) so that survey results showed that 'innovating businesses in London were externally oriented, and with no particular interest in local clustering' (Buck et al. 2002: 120). So London's economic advantages – just as much as Johannesburg's – are located in a very generalised set of agglomeration economies that 'presumes no form of co-operation between actors beyond what is in their individual interests in an atomised and competitive environment. The system is without any particular observable organisation or inter-agent loyalty, and simply functions as an ecology of activities benefiting from proximity' (Gordon and McCann 2000: 517). Firms and economic actors of various kinds and sizes, in diverse locations, share the physical fabric of the city and the generalised benefits of co-location. In one of the world's largest and most knowledge-based urban economies, then, the value of the wider city to the economy (in terms of pure agglomeration economies and a widely accessible infrastructure to support circulation and interaction) is strongly evidenced.

All this suggests the importance of generating policy recommendations and political arguments for a quite different range of interventions than those that support specialised clusters – interventions that are perhaps more appropriate to cities with limited skills bases and few technology-dependent activities, with strong external dependencies, very diverse economic activities and scarce public resources. These would be interventions that take seriously the positive benefits of generalised urban externalities and that would aim to build on the potential resilience of diverse economies, rather than bemoan the absence of specialised or globalising clusters (Douglass 2001). Such advice could work equally as effectively in Johannesburg as in London – and could be in direct opposition to recent policy initiatives in both cities that have tended to prioritise global economic sectors as drivers of future development (Buck et al. 2002, Gordon 2003, Parnell and Robinson 2005).

A first step here would be to consider some existing theoretical arguments about how diversity might work to expand urban economies and also to support innovation. An early argument to this effect was put forward by Jane Jacobs, whose vision of the socially diverse, ordinary city also informed her assessment of how cities grow economically. As she noted, 'In cities, liveliness and variety attract more liveliness; deadness and monotony repel life. And this is a principle vital not only to the ways cities behave socially, but also to the ways they behave economically' (Jacobs 1965: 109). In her view, the vitality of the city, its ability to function well and to be generative of social and economic innovations, depended to a large extent on the co-presence of a diversity of people and a variety of different uses, physically in specific city streets and neighbourhoods, but also across the city as a whole. While her arguments concerning the social uses of 'disorder' (as Richard Sennett (1990) has famously argued) have become well known, the economic uses of diversity have been explored less.

Jacobs's overall argument is that it is the everyday, tolerant coexistence of social difference and economic diversity that enables cities to bring together different people and different talents, to generate new activities and to enable innovation. Cities that encourage segregation – physical or social – from the 'ordinary' city, with its jumbled mixtures of uses and people, she suggests, will lack vitality and perhaps die. Such predictions have certainly not come true – places she cites as dysfunctional, such as Los Angeles and Boston (1965: 83–4) have instead become home to some of the most rapid urban economic growth of the last decades of the twentieth century (Markussen and Gwiasda 1994). And this is perhaps because in her enthusiasm for a specific urban spatial form – the socially diverse, mixed-use neighbourhood – the wider argument about the virtues of the coexistence of different economic activities in large cities has sometimes been obscured. Clearly, diverse neighbourhoods, the physical coexistence of a range of business environments and the spatial concentration of diverse activities can have the generative social and economic effects she describes. But cities do not always get to be like that, and yet they remain sites for economic growth and innovation.

Cities enable opportunities for frequent interactions and sustained relationships amongst economic agents and social groups, both within and even across 'segregated' spaces and relationships. The benefits of proximity for networks of learning, cooperation and competition in dynamic clusters and specialised concentrations in some cities would be very relevant examples of innovation through single-use neighbourhoods, a kind of 'segregation'. But, in many cases, strong concentrations of certain kinds of activities in cities do not translate into close spatial clustering, or into close interactions amongst similar firms, or even into stable ongoing interactions with actors involved in supplying services, inputs or marketing support (as Buck et al. 2002 show in the case of London). Thus, and especially in a world of distanciated social relations where relational proximity is an increasingly important basis for learning (Amin 2003), we can expect that innovation will survive efforts by various actors to 'decontaminate [themselves] from the ordinary city' (Jacobs 1965: 180). So, in economic terms, Jane Jacobs's 'ordinary city' is perhaps best thought about at a city-wide scale where the diversity of activities that she identified as crucial to economic expansion and innovation coexist and have opportunities for the interactions she saw as crucial for generating new activities.

This is the hunch of some contemporary economic theorists, who suggest that economic diversity might be a special virtue of larger cities and that it is this diversity that enables new kinds of activities and new ways of doing things. Duranton and Puga (2001) argue that 'nursery cities' – large, diverse urban centres – are an important feature of the landscape of cities and explain the economic dynamism of cities as much as spatially concentrated specialisations. They suggest that 'Most innovations take place in particularly diversified cities and most new plants are created there' (2000: 538). Working through formal economic theoretical assumptions about the costs and benefits of urban agglomeration, they introduce a stronger role for shared local inputs across sectors, especially during phases of innovation and experimentation in the lifecycle of firms. Economic diversity, and an appreciation of both localisation (sectorally specific) and urbanisation (cross-sector) economies are important here. These dynamics contribute to the emergence of both specialised and diversified cities in city systems or the coexistence of specialisation and diversity in large cities.

They identify 'dynamic advantages based on diversity [. . .] because a city with a wider range of local sectors using different processes allows firms producing new products to try more processes in search of their ideal one without incurring relocation costs' (2000: 551). Large, diversified cities function as 'nursery' cities, enabling firms in an innovative and learning phase to benefit from the range of different economic processes and actors available as well as from shared pools of skills and other resources. Since 'not every community can create the next Silicon Valley in its backyard' (2000: 553), the assessment that large, diverse cities play an important role in economic innovation and expansion has strong policy implications. As they comment, 'the need for (and the type of) intervention in the presence of localisation and urbanisation economies depends delicately on the source of these economies – something that is far from well understood' (2000: 552–3). Specialisation, although not ubiquitous in cities, of course contributes to the dynamism of urban economies. But if the ambition is to encourage economic growth, then the uses of economic diversity also need to be appreciated.

Supporting Johannesburg's diverse economy

Policy-makers in Johannesburg, confronted with the ordinary city and a diverse economy, have been mindful of the importance of building the broader agglomeration economies of the city. Policy statements have proposed initiatives to enable circulation and to improve transport connections, have sought to address crime as a key disincentive for investment and to build the broad educated skills base needed for future development across a diversity of economic activities. They have also brought into focus a key industry – the logistics sector – which plays an important role in the effective operation of many others. And there were significant reasons to continue to support even declining manufacturing sectors because they offered a strong employment base for relatively unskilled labour.

However, if Duranton and Puga and Jane Jacobs are correct, there are grounds for a more strategic appreciation of economic diversity as a potential basis for economic growth. In addition to the potential role of diversity in fostering innovation that these authors identify, some of the activities that play a prominent role in Johannesburg and other large cities depend precisely upon the existence of a diverse economic environment for their operations. Henderson (2002: 98), for example, suggests that these include production and retail of high-tech products, research and development activities, financial services, company headquarters and business services. In South Africa these activities are all significantly concentrated in Johannesburg. Using 1999 data, Rogerson (2001) has calculated that 73 per cent of South Africa's high-technology manufacturing employment and 78 per cent of IT service-sector employment is concentrated in the broader region of Gauteng, of which the Greater Johannesburg Metropolitan area comprises about 70%. Although the local market was crucial for high-tech manufacturing firms,[12] Rogerson notes that 'Many of these companies [. . .] identified as a strength their linkages or membership of large corporate groups (often overseas linked) which afforded access to new technologies, markets, finance as well as exposure to international quality standards' (2001: 43). Support for Johannesburg's economy needs to foster these external relationships as much as it needs to focus on building local social relationships.

Some of the key influences on firms' choices to locate in Johannesburg, despite frequent acknowledgement of the better quality of life in Cape Town, included the quality of infrastructure (international airport and roads), the large supply of skilled labour and the centrality of the city in terms of the rest of the country's markets. So, although Rogerson concludes that 'Gauteng shows the features of what, in other countries has been called an innovative milieu' (2001: 46), the evidence so far suggests that it is at the city-wide scale, through the broadest agglomeration economies (of skills, markets, infrastructure) and through wider external connections, that opportunities for expansion and innovation have been generated.

The value of appreciating the diversity of urban economies has added relevance in contexts with substantial levels of informality in economic practices. Here the redistributive consequences of attending to diverse economies and broader agglomeration economies are potentially very important. Economic policies that focus on clusters and specialisation can certainly incorporate informal sector firms, perhaps through appreciation of clustering practices amongst these firms themselves (Rogerson 2002). But evidence from a substantial study conducted under the auspices of the World Bank in Johannesburg in 1999,[13] a survey of about 500 informal firms, just over 300 large manufacturing firms, and over 800 small, medium and micro-Enterprises (SMMEs), demonstrated the importance of wider inter-connections amongst different kinds of local firms. Forward and backward linkages amongst firms within the city-region were substantial, including between large firms and SMMEs and informal firms. Between 20 per cent and 48 per cent of large firms reported that they drew on informal and small businesses for inputs. As the report notes, 'Of all LFS [Large Firms Sector] firms, as many as 48 per cent buy their semi-processed inputs, 39 per cent buy their professional services, and 20 per cent buy their shipping services from the small and informal firms tier' Chandra et al. (2001: 12–14).

The potential for increasing forward and backward linkages from informal sector firms,[14] though, was shown to be severely limited by the lack of access to finance and the spatial segregation of largely black informal sector operators (mostly based in former black townships and inner-city areas) from formal large firms and from wealthier customers (Chandra et al. 2002). Although the apartheid-derived spatial mismatch between the concentrations of major employers and the poorest residents of the city didn't seem to adversely affect larger firms, for smaller and informal firms, almost half of whom were based in the home, these locational disadvantages were a serious problem. Lack of access to services and the difficulties of public transport meant that substantial support for informal business activities could be achieved by upgrading poorer residential environments. As the authors of the report note, 'The challenge of upgrading informal business districts thus becomes inextricably linked with the upgrading of poor residential areas, making it difficult for local authorities to target informal business locations without addressing the larger problem of residential township upgrading to improve infrastructure and service delivery' (Chandra et al. 2002: 16). Since there is clearly scope for interactions with larger firms in terms of both inputs and outputs, policy initiatives might attend to broader infrastructural provision for the city to encourage stronger opportunities for interaction amongst firms and circulation across the city. This could also open up new kinds of demand for goods and services, which was identified as a key constraint on the growth of SMMEs and informal sector activities.

Scott and Storper (2003: 588) suggest that although basic infrastructure is a prerequisite for successful economic development, the focus of policy attention should be on facilitating the development of the endogenous social and cultural assets of urban agglomerations. One concern might be that the high cost and targeted nature of strategies to support social and cultural assets in a city have the potential to restrict attention once again to clusters and formal activities in the wealthiest parts of cities. However, their interest is in enhancing the 'regional economic commons', which reinforces the argument being made here for supporting the broadest agglomeration economies of cities. And here, enhancing and facilitating the infrastructures and circulation of people, ideas and goods in and beyond cities might ensure that public funds are directed to the widest possible set of economic activities, supporting high-tech global firms as well as the smallest informal operator.

In the case of Johannesburg, there would be an argument for placing this assessment about the benefits of economic diversity within the broader city-region. For, although the Unicity process described in Chapter 5 brought together a racially and administratively divided city, the functional economic region cuts across a number of contiguous and similarly large municipalities (Pillay 2004, South African Cities Network 2004; see Fig. 5.2, Map C). The Greater Johannesburg Municipal Council area has certainly seen declining manufacturing production and a rise in employment in skilled and high-tech sectors. But as manufacturing relocates out of the municipality, it is mostly moving to neighbouring parts of the Gauteng region (Rogerson and Rogerson 1999). This movement within a broader city-region has also been identified in cities such as São Paolo and London (Rodriguez-Posé et al. 2001, Buck et al. 2002). Still reliant upon the broad agglomeration economies of the city, new and more suitable business environments can be sought in the wider city-region. Contrary to Jane Jacobs's often very small-scale analysis, then, the 'ordinary city' in all its diversity and multiplicity is perhaps best observed at the city-region scale, where interaction and circulation, access to a diversity of skills, resources, labour and sectoral and cross-sectoral learning is possible across the region.

The diverse economic activities that coexist in the city rely, then, on the common broader inputs of the city infrastructure, shared labour resources and skills, the wide range of locally available inputs and interactions with different firms as well as access to regional and international locations offered by communications networks. Firms of various sizes and from various sectors benefit from co-location, sometimes from specific interactions and linkages with closely related firms, others from the potential to engage at some stage with a wide range of other firms and more generally from the shared institutional and infrastructural environment within which they function. The operations of the largest international firm and those of the smallest survivalist trader draw on the city's resources, potentially benefit from the collective infrastructure of the city and depend on the presence of other firms and many different residents in order to maintain and perhaps expand their activities. The fabric and form of cities, the activities they assemble and the collective social life they foster all contribute to, sustaining and sometimes thwarting the activities of these diverse enterprises. Policies that aim to enhance these wider agglomeration economies have the potential not only to support economic growth through diversity, however, but also to address the challenges of the wider social and political context of cities. This could be crucial in resource-poor cities but, as we have seen, it is also important in wealthier city contexts.

CONCLUSION

This chapter has considered the consequences for urban policy of seeing all cities as ordinary. Some new insights emerge from seeing cities as diverse social and economic assemblages in which visions for the future of cities are certainly politically contested, but where the range of activities, interests and wider connections can also be the foundation for economic expansion and building better urban futures. Whereas much policy advice has stressed the potential of stimulating strongly localised learning environments through supporting specialised clusters, this ignores the vast potential that general urban agglomeration economies and the presence of a diversity of economic activities in cities might have for stimulating sustainable, resilient and even expansive development.

These conclusions certainly matter, as we have shown, for poorer, unequal cities that, like Johannesburg, could find strong synergies between generalised service delivery, infrastructure development and economic growth. They indicate directions for reflecting on how even very poor cities might embark on useful initiatives to support and hopefully expand the diverse activities that provide incomes and sustain livelihoods in different contexts across the city. Finding a job or generating an income are the top priorities of the poorest people in Johannesburg. For poorer cities, supporting economic growth and expansion is often an important policy agenda. But so many of the strategies for achieving this are imported from wealthier city contexts, where the resources and opportunities for developing urban economies have been somewhat different. This chapter has shown how keeping in mind the ordinary city – diverse, contested, connected – brings into view a range of potential strategies for enhancing economic growth that could encourage redistributive investments in infrastructure, as well as expand access to employment and support different kinds of livelihoods. Reinforcing the broadest agglomeration economies of cities, it has been suggested, along with encouraging diverse urban economies, might be more useful strategies than pouring scarce public resources into already successful clusters of globalising firms.

This analysis is also relevant to wealthier cities, as we saw in the case of London (Gordon 1999). Indeed, the futures of cities such as New York and London might depend on enhancing and supporting a diversity of economic activities and retaining a multiplicity of functions in order to avoid the kind of risky dependence on a small number of economic sectors that could experience a generalised decline or find locational parameters changing, with severe consequences for these cities. As Markusen and Gwiasda note, 'In our view, it is a mistake to celebrate New York's increasing dependence on internationally oriented producer services' (1994: 185). Successive develop-ment regimes there have selectively enhanced conditions for the rise of these activities and have hastened the decline of manufacturing industry (Abu Lughod 1999, Sites 2003). Along with post-financial crash east Asian cities, where the need for resilience and longer-term security through economic diversity has been starkly demonstrated (Firman 1999, 2000 Douglass 2001), urban managers in wealthier contexts can certainly benefit from thinking of their cities as ordinary.

In this regard, there is much to learn from attempts to imagine a role for urban governance in contexts of substantial informality and poverty. It has been argued – although not often implemented – that, if resources permit, the best interventions in contexts where much activity needs to escape the surveillance of the state to succeed might be to follow the inventive activities of urban dwellers, as they stitch together livelihoods and opportunities across and with the city. Supportive, facilitative infrastructure development and permissive, enabling and participative service delivery could build diverse, ordinary cities alongside and for urban residents (Simone 1998, Gotz and Simone 2003). As Amis (2002: 109) notes:

> Municipalities should concentrate on their traditional roles of providing infrastructure, ensuring health and education, and appropriate planning and regulation. Despite the hype about city marketing and mega projects, traditional service delivery may be more critical. For policy makers this means concentrating on not destroying employment opportunities as well as undertaking traditional functions as efficiently and equitably as possible. The poor benefit disproportionately from the efficient and inclusive delivery of services.

Refocusing policy interventions in wealthier contexts in the tracks of govern-ance initiatives in some of the world's poorest cities suggests a stronger focus on the broadest agglomeration economies of cities (see also Turok et al. 2004) and room for exploring the economic significance of basic infrastructure development and effective service delivery to foster a diverse, dynamic urban economy.

Urban theory has not always been helpful in directing our attention to the potential of either diverse economies or ordinary cities. Recent versions of hierarchical thinking within urban theory – ironically in accounts of globalisation and world-city networks – have left many poorer cities tossed between ambitions to globalise and the demands of developmentalist policy. As we saw in Chapters 4 and 5, appreciating cities as ordinary, rather than ascribing to them a position within a spurious hierarchy based on a few restricted and contentious criteria, offers the potential to reframe the policy and political challenges of imagining city futures.

Developing visions for the economic futures of ordinary cities, then, requires theoretical repertoires that are appreciative of the diversity of city economies and that aim to work beyond hierarchies, beyond a divide between globalisation and development and beyond an ascription of inventiveness and modernity only to wealthy cities. Bringing the (ordinary) city back in places the challenge of visioning city futures within a more cosmopolitan field of research and policy. This is not to suggest that such approaches to the challenge of cities can escape the fields of power that frame urban policy and urban theory, as the case study of Johannesburg's city-visioning process amply demonstrates. The futures of cities are sites of contestation: much is at stake in the analyses that inform debates about these futures as well as in the deeply contested politics that will shape them.

Conclusion

Caught between accounts of modernity and development, which have both constrained understandings of cityness and limited expectations of urban futures, urban studies is ready for revitalisation. More than this, facing a world of ever-expanding cities, more likely than ever before to be home to poor people, urban studies cannot continue to base its theoretical insights on the experiences of a few wealthy cities: this would doom it to irrelevance. To address the twin problems of ethnocentrism and intellectual division, I have articulated a post-colonial account of ordinary cities and proposed that from the ruins of an urban theory divided by colonial pasts and developmentalist assumptions might emerge an understanding of all cities as ordinary. Ordinary cities, I have suggested, are distinctive and unique, yet they are all potentially part of the same field of analysis. Certainly, a different understanding of cityness has come into view as a result of the extended post-colonial critique offered here. Each chapter in this book has experimented with tactics that instigate a postcolonialisation of urban studies; each chapter has contributed something to the conceptualisation of ordinary cities; and most of the chapters have suggested some implications of ordinary cities for urban policy. In bringing the discussion to a close it will be helpful to reflect on these three issues that have been running through the different chapters: How to go about doing post-colonial urban studies? What are ordinary cities like? And what might urban policies for ordinary cities look like?

TACTICS FOR A POST-COLONIAL URBANISM

It's easy enough to raise a critique of ethnocentrism in urban studies; it's so much harder to instigate new kinds of practices for studying cities and managing them. It can take a lifetime to learn about one place – how much more difficult to build analyses through a range of different contexts! This book is exemplary in that: I have managed to learn more about cities in the UK and in the USA (which those of us working on poorer contexts generally know something about, since that is where 'the literature' has conventionally been produced). I have had some help in learning something about a few cities in Brazil, Malaysia and Zambia. And I have my own experience of South African cities to draw on. So many places remain to be considered, so many different cities still need to be drawn into the task of exposing and then critiquing the assumptions on which wider accounts of cities depend. My hope is that scholars from different contexts will be able to identify with the broad challenge to postcolonialise urban studies and take the analysis further.

The call to postcolonialise our research can appear daunting and even overwhelming for scholars and practitioners who are all too busy engaging with the rich complexity of their own contexts. My point, however, is definitely *not* that urban researchers now need to rush off and study many different contexts before they can contribute to a postcolonialised urban studies. In fact, the opposite is closer to my personal position: I would be horrified if the consequence of this book was that well-resourced scholars went globe-trotting, in the wake of their late-colonial predecessors, to study cities around the world. Some people might: and my hope is that those of us who live and work in discrepant, often diasporic relationship to the places we research will be responsible, engaged and supportive members of intellectual communities concerned with urban processes in those places.

My ambitions for urban scholarship are rather more modest. They include the hope that we will find ways to operationalise Dipesh Chakrabarty's desire to 'provincialise Europe' (2000). He comments that in the shadow of colonialism, 'Europe works as a silent referent' in much scholarship. One of the consequences he observes in the field of history – equally valid for the field of urban studies – is that while 'Third-world historians feel a need to refer to works in European history; historians of Europe do not feel any need to reciprocate' (2000: 28). He continues to suggest that:

> 'They' produce their work in relative ignorance of non-Western histories, and this does not seem to affect the quality of their work. This is a gesture, however, that 'we' cannot return. We cannot even afford an equality or symmetry of ignorance at this level without taking the risk of appearing 'old fashioned' or 'outdated'.

So, if I was allowed to name the most promising tactic for post-colonial scholarship, I would suggest that any research on cities needs to be undertaken in a spirit of attentiveness to the possibility that cities elsewhere might perhaps be different and shed stronger light on the processes being studied. The potential to learn from other contexts, other cities, would need to always be kept open and hopefully acted upon. In this sense Chakrabarty's aphorism – Provincialise Europe! – is a little misleading, for he is asking that we open up Western scholarship to insights and comparisons from elsewhere. Really, we should be making European (and American, South African, Malaysian, etc.) scholarship more cosmopolitan in their sources of inspiration and learning. The primary tactic that I propose for postcolonialising urban studies, then, is one of decentring the reference points for international scholarship. Thinking about cities ought to be willing to travel widely, tracking the diverse circulations that shape cities and thinking across both similarities and differences amongst cities, in search of understandings of the many different ways of urban life.

However, there are powerful structures that shape the production and circulation of knowledge about cities, starting with the resources to research and reflect on cities and extending to the publishing business as well as the nature of the market for books and articles (Robinson 2003b). The resources for research are very unevenly distributed, and the publishing industry strongly privileges the markets of wealthy countries in making decisions about what sorts of books to publish. Journal editors and referees likewise reinforce a certain privileging of some parochial knowledges by insisting that writers on contexts outside the West refer to the overwhelmingly Western literature in developing their arguments. Urban studies is in the fortunate position of having some core journals committed to publishing work from many different contexts, but it is possible to give more scope to accounts of cities that decentre dominant literatures. This has the potential to challenge urban theories that parade as universal while being in reality rather parochial. This is an exciting, if difficult, project, but it is also a very practical tactic for promoting a post-colonial urban studies.

Through the various chapters in this book I have experimented with a few different intellectual strategies for postcolonialising understandings of cities. I started by turning my attention to Western accounts of urban modernity, with Chakrabarty's injunction in mind, to provincialise these understandings of cities. So as a first cut at postcolonialising urban studies I determined to locate the places that apparently universal accounts of cities had in mind.[1] Where were the cities that informed dominant understandings of cityness? The second tactic I found helpful was to dislocate the association between understandings of cityness and those privileged sites for theoretical reflection. The geographical referents for urban studies, I suggested, needed to be diversified. In doing this I was inspired by the comparative approaches of an earlier generation of scholars who drew on studies of many different cities around the world to critique the ethnocentric theories of urbanism that dominated the field. Attending to the differences between cities was most helpful for parochialising apparently universal accounts and for insisting that ways of urban life were perhaps as diverse as the cities they inhabit.

But comparative accounts have their limitations. They focus our attention on tracing similarities and differences and can too easily reinstate categorisations of cities or rely on implicit assumptions about convergence amongst cities. Perhaps more persuasive as a tactic for postcolonialising urban studies would be taking a cosmopolitan perspective on urban processes and experiences. Here I was inspired by James Clifford's (1997) analysis of the diverse, or 'discrepant', cosmopolitanisms of most societies. Far from proposing a new kind of Kantian universalism or a transcendent account of cityness,[2] this cosmopolitanism evokes the diverse trajectories of people, resources and ideas that make up cities. It draws our attention to cities as distinctive assemblages of many different kinds of activities and helpfully disaggregates any comparative exercises: cities are many things, and following the trajectories of different elements of any urban context quickly brings other cities into view in quite specific ways. Ordinary cities, then, are distinctive and have the capacity to shape their own futures, even if they exist in a world of (power-laden) connections and circulations. Deploying a cosmopolitan analytic dispelled any sense that some cities are originators or exemplars and others imitating, or backward. This had important consequences for accounts of urban modernity and also for understandings of urban development. In both cases, being alert to the cosmopolitan nature of cities illuminated the ways in which borrowing, adaptation and invention all contributed to making distinctive cities.

These tactics also informed the assessment in Chapters 5 and 6 of what kinds of strategies for urban development might be relevant for cities within a post-colonial perspective. Learning from the experiences of different cities and the interventions inspired by different contexts demonstrated that urban scholarship need not be divided by assumptions about levels of development or assigned position in any proposed hierarchy. Initially, I adopted the tactic of bringing the divided scholarships of poorer and wealthier cities together in an explicit encounter between literatures concerned with globalisation (in Chapter 4) and those concerned with development (in Chapter 5). The final chapter initiated a more cosmopolitan exploration of urban development that drew on a range of different cities. The suggestion, then, is that all cities can be considered within the same field of analysis, attentive to the diversity of urban experiences and the range of challenges that face cities in different contexts.[3]

THE DIVERSE SPATIALITIES OF ORDINARY CITIES

The tactics that I have used to provoke a post-colonial critique of urban studies have suggested the need for a new account of cities – one that treats all cities as ordinary. For me, the label 'ordinary cities' applies to all cities: the wealthiest cities and the poorest, those that are host to the more powerful organisations and agents in a globalising world and those where residents have very little obvious capacity to shape changes in the broadest structures of the global economy. This operationalises the fundamental requirement for a post-colonial urbanism: the possibility to consider all cities within the same field of analysis. Poor cities should not drop off the radar of what it means to be urban just because poor people live there, or because infrastructures are crumbling, or because informal processes determine much of political and economic life. And wealthy cities should not get to determine what it means to be urban, just because they and the scholars who live there have the resources to crowd out analyses of other cities. It has been a consistent theme of this book that all cities, from Los Angeles to Lagos, from Johannesburg to Jakarta, would be better off for being understood as ordinary. This is not to suggest they are uninteresting, or to imply that cities are best explored through the mundane and banal activities that we might think of as unexceptional – although these are certainly important to understanding what cities mean for social, economic and political life. Instead it is to insist that all cities are distinctive and unique rather than exemplars of any category: indeed, for those who live in them, very often individual cities are special and particularly meaningful places.

So the broadest implication of seeing all cities as ordinary is that what gets to count as 'urban' needs to be informed by an engagement with the diversity of cities and the diversity of ways of living in cities. Bringing all cities within the same field of analysis through the idea of ordinary cities will ensure that no particular cities or group of cities will a-priori determine how cityness is represented.

But the analysis in this book has also developed some more substantive understandings of what ordinary cities might be like, at least partly as a result of the postcolonialising manoeuvres adopted. In a post-colonial spirit, though, these must necessarily be always open to revision, the meanings of cityness perhaps indefinitely deferred as cities inventively transform themselves and scholars reflect the diversity of urban experiences against extant analyses. Most especially, the substantive conclusions of this book have concerned the spatial imaginations that can help us to appreciate ordinary cities. Building from a critique of categorising and hierarchising accounts of cities I have suggested that to appreciate ordinary cities requires attention to a diversity of spatialities. Globalising and developmentalist accounts of cities drew attention to only some localised parts of cities – globally connected clusters or infrastructurally poor locales. As Amin and Graham (1997) observe, these accounts draw on the metaphorical tactic of synecdoche (letting the part stand in for the whole) to represent cities through only partial aspects of their incredible diversity. And, although an enthusiasm for tracing networks and connections amongst cities has been very productive for appreciating how cities work through the wider flows and interaction that they enable, these analyses have been too restricted in terms of the connections and networks that matter to them. Amin and Graham propose instead that a much more complex spatial imagination would help us to better appreciate ordinary cities, in all their diversity. They hope that urban theory might:

> overcome the limits of partial perspectives and its tendency to rely on paradigmatic cases. It would enable more subtle perspectives on urban multiplicity stressing the interconnections between the complex time-space circuits and dimensions of urban life, as well as the diversity and contingency of the urban world.
>
> (1997: 420–1)

They see cities as 'multiplex' spaces, sites where many different overlapping networks of association and interaction come together, leading to inter-actions but also to the fragmentation of city spaces and the potential for disconnection. This is a sophisticated sense of the spatiality of cities, one that is also invoked by Amin and Thrift (2002), Graham and Marvin (2001) and Massey, Allen and Pile (1999). It chimes with the conclusions of the post-colonial analysis offered here in which the diversity of ways of city life are foregrounded, including the relations of interaction and disconnection, sociability and alienation that Chapter 2 explored. Moreover, focusing on the wide range of circulations and connections amongst cities means that even apparently quite different cities can be thought of as already connected and influencing one another.

However, analyses that pay closer attention to the diversity of cities might press at the edges of this spatial imagination by stressing the crucial signifi-cance of the territory of the city. Bringing the city back in, as I argued in Chapter 5, further diversifies the spatial imagination of urban analyses and captures some important aspects of ordinary cities. These include illuminat-ing the ways in which cities operate as platforms facilitating diverse economic activities, as sites for redistribution and as arenas for political contestation. The need to reterritorialise the spatial imagination of urban studies emerges from this post-colonial account of ordinary cities. I very much hope, though, that the diverse spatialities of ordinary cities will be explored by scholars of cities that I have not learnt from in this book, or by investigators excavating aspects of the complex life of cities that I have not been able to attend to here. For it is in the nature of ordinary cities that their multiplicity and complexity will always escape us, and in a world of cities there will always be much to learn.

LETTING CITIES BE ORDINARY

I have argued that until recently urban policy, like urban theory, has been very segmented between wealthier and poorer cities. And that even now many policy interventions are framed through the lenses of developmentalism and globalisation. Reinforced by long-standing cultural imaginaries of some cities as mere imitators of others, the domain of urban policy, as much as that of urban theory, could benefit from an appreciation of cities as ordinary. Releasing cities from these colonial assumptions about their modernity has the potential to enable city managers and residents to frame policies that address the distinctive needs of their city.

However, the terrain of urban policy has also harnessed the political ambitions of urban managers and elected officials to the hopes of com-petitive success, especially moving up the hierarchy of cities through investing in globalising economic activities. Letting cities be ordinary would be an important foundation for a different set of policy initiatives. Such initiatives, we have seen, would attend to the diversity of social life and economic activities in cities. They would avoid making a-priori assumptions about the kinds of activities – either in terms of sector, reach or formality – that would lead to economic growth in any particular city. We have also seen that it is precisely the diversity of ordinary-city economies that supports economic expansion and innovation, and we have seen that policies that aim to support this diversity through enhancing the widest agglomeration economies of the city would also be well placed to support social needs and to address political demands from across the city.

It is the strongest conclusion of this book that the field of urban studies and the task of improving life in cities will be vastly enhanced by letting cities be thought of as ordinary. This will involve a continuing critical evaluation of the field's own complicity in propagating certain limited views of cities and thereby undermining the potential to creatively imagine a range of alternative urban futures. It will require more cosmopolitan trajectories in assembling the sources and resources of urban theory. Much innovative work is already being undertaken by scholars and policy-makers around the world who have had to grapple with the multiplicity, diversity and ordinariness of their cities for some time. Ordinary cities are themselves enabling new kinds of urban imaginaries to emerge – it is time urban studies caught up. More than that, I would suggest that as a community of scholars we have a responsibility to let cities be ordinary.

Notes

INTRODUCTION: POST-COLONIALISING URBAN STUDIES

1 Previous contributors to this endeavour include Southall (1973), Lawson and Klak (1993), Ward (1993), King (1995) and Simon (1995).

1 DISLOCATING MODERNITY

1 See Crysler 2003, Chapter 4 for some recent debates about this within the field of architecture.

2

> [M]odernity is not, as such, a project. It is a form of historical conscious-ness, an abstract temporal structure which, in totalising history from the standpoint of an ever-vanishing, ever-present, embraces a conflicting plurality of projects, a conflicting plurality of possible futures, provided they conform to its basic logical structure. Which of these projects will turn out to have been *truly* modern, only time (historical time) will tell.
>
> (Osborne 1992: 37)

3 As Anthony King notes, within urban scholarship modernity and tradition have been relationally defined – as with other binary concepts, each makes the other possible (2004: 72).

4 In his critique of the Chicago School, Castells (1977: 17) criticised the ideological nature of the definition of the urban proposed by them, writing, 'Indeed, the impossibility of finding an empirical criterion for the definition of the urban is merely the expression of theoretical imprecision. This imprecision is ideologically necessary in order to connote, through a material organisation, the myth of modernity.'

5 In his essay on surrealism Benjamin observes that Breton 'was the first to perceive the revolutionary energies that appear in the "outmoded"' (1979: 229).

6 'Just as the living room reappears on the street, with chairs, hearth, and altar, so, only much more loudly, the street migrates into the living room' (Benjamin 1979: 174).

7 So rather than seeing these essays on Naples and Moscow as a seamless part of Benjamin's analysis of 'city life' as Savage (1995: 205) suggests, I want to argue that these writings reflect an engagement with the specific differences he encountered in these cities.

8 Naples: midday sleep 'is not the protected Northern sleep. Here, too, there is interpenetration of day and night, noise and peace, outer light and inner darkness, street and home'. (Benjamin 1979: 175). 'Poverty has brought about a stretching

of frontiers that mirrors the most radiant freedom of thought. There is no hour, often no place, for sleeping and eating' (1979: 175).

9 A feeling for the valuation of time, notwithstanding all 'rationalization', is not met with even in the capital of Russia. [. . .] From earliest times a large number of clockmakers have been settled in Moscow. They are crowded, in the manner of medieval guilds, in particular streets, on the Kuznetsky Bridge, on Ulitsa Gertsena. One wonders who actually needs them. 'Time is money' – for this astonishing statement posters claim the authority of Lenin, so alien is the idea to the Russians. (1979: 189)

10 'And something else, too, reminds one of the South. It is the wild variety of the street trade [. . .] all this sprawls on the open street, as if it were not twenty-five degrees below zero but high Neapolitan summer' (1979: 180–1).

2 RE-IMAGINING THE CITY THROUGH COMPARATIVE URBANISM: ON (NOT) BEING BLASÉ

1 This includes the category of the colonial city which, in insisting on the importance of the colonial experience for cities in former colonies, hides from view so many of the aspects of urbanism in these cities that either had little to do with colonialism or that represented distinctive local combinations of what were very common circulating urban practices and built forms, albeit under the political aegis and sign of colonial power. The Zambian cities we will be exploring were indeed cities forged under colonialism and, while colonial power relations were relevant, as we will see, to the urban ways of life that emerged there, it would be quite inappropriate to see all events and processes there through that lens. As King (1990) notes, there is an argument for thinking of the wealthy cities of colonial powers as 'colonial' cities too.

2 His ecological predilections (what contemporary theorists might describe as the causal aspects of urban space) inclined him to consider the city as generative of forms of social life (Sjoberg 1959, Castells 1977), although external and more properly sociological factors also figured in his analysis.

3 Definition from the *Oxford English Dictionary*, 'debility of nerves causing fatigue, listlessness, etc.'

4 This was a common strategy for these anthropologists, working in the colonially divided towns of Zambia. Schumaker notes that Epstein had wanted to live in the township amongst African people but colonial restrictions forbade this. Instead, 'he placed his research assistants and their wives in different parts of the town and set them to observing urban life. He spent his time teaching the men to write reports and to produce texts on union and political meetings and to conduct informal interviews'. He himself attended many political and union meetings and befriended key officials (Schumaker 2001: 181). See also Powdermaker (1966) for some personal reflections on this.

5 Most of the research reported here was undertaken when 'Zambia' was known as Northern Rhodesia, a British colony. For simplicity's sake and obvious political reasons, I refer to the country throughout by its post-independence name of Zambia.

6 Gluckman is following Mitchell here, who notes: 'their own ethnic distinctiveness which they took for granted in the rural areas is immediately thrown into relief by the multiplicity of tribes with whom they are cast into association. Its importance to them is thus exaggerated and it becomes the basis on which they interact with all strangers' (Mitchell 1956: 29).

7 These three languages belong to the same linguistic group (Mitchell's footnote.)

8 Lamba are the chiefdoms adjacent to the town (Mitchell's footnote.)

9 He uses the Anglicism sipiti = speed (Mitchell's footnote.)

10 That the authorities disagreed with some of the writings of these anthropologists and often chose to ignore them has been well documented (Hannerz 1980: 157–60, Schumaker 2001) – although so have their blind spots and colonial assumptions (Magubane 1968, Ferguson 1999).

11 Although some recent comparative studies of this nature include Lee (2001) and Oliviera (1996).

12 Oscar Lewis puts it this way:

> Anthropologists have a new function in the modern world: to serve as students and reporters of the great mass of peasants and urban dwellers of the underdeveloped countries who constitute almost 80 per cent of the world's population. What happens to the people of these countries will affect, directly or indirectly, our own lives. Yet we know surprisingly little about them.
>
> (Lewis 1959: 1)

13 This is a question Michael Cohen (1997) asks too, and answers by suggesting that the developmentalist concerns of southern cities are broadly applicable to all cities, although he suggests they are similar rather than identical.

3 WAYS OF BEING MODERN: TOWARDS A COSMOPOLITAN URBAN STUDIES

1 Le Corbusier is quoted as observing that it was 'In New York, then, I learned to appreciate the Italian Renaissance' (Taylor 1992: 63).

2 The term that Cardoso (2003) uses to characterise Latin American architecture.

3 And indeed to this day 'the topic of imitation and originality (is) a constant preoccupation of literary histories in Latin America' (Kirkpatrick 2000: 196).

4 From Comaroff and Comaroff (1997: 233), who draw the proverb from a compilation by Sol T. Plaatje, African writer and historian, published in 1916. In Tswana it reads, *Loare go bona sesha lo se eka-eke lo latlhe segologolo sa lona*.

5 Comaroff and Comaroff (1997: 251) discuss the ways in which through 'refined consumption', better-off African workers in the cities of South Africa and in the countryside might use clothing styles to 'contrive respectability and personal distinction in urban colonial society'. In the southern African context more widely, many urban dwellers and migrants to cities direct a lot of care and attention to clothing, and events that stage competitions for best-dressed entrants, as well as song and dance contests that include smart outfits, are popular.

4 WORLD CITIES, OR A WORLD OF ORDINARY CITIES?

1 There are many problems associated with assigning 'power' to the category of global city, when the capacities to control and command that are being identified are located in certain actors and institutions within a small part of the city's economy. 'Cities' as collective actors, may in fact be rendered relatively powerless in these contexts. Moreover, as John Allen (1999) astutely asks, it is unclear where power is produced – in the sites Sassen identifies – the city neighbourhoods, or in the network interactions and flows. Taylor et al. (2001) are beginning to explore the idea of networked power. But the thinking that elides categories of cities with economic agents in relation to power continues to have purchase and, as I will argue below, important effects.

2 From Beaverstock et al. (1999: 457).

3 Taylor (2004) identifies Lusaka as part of a cluster of inter-tropical African cities that, on his index of service sector-based connectivity, are 'lowly and relatively isolated'. Lusaka is, however, considered here, alongside another 192 cities that fall outside the 123 'top' world cities, for the first time by Taylor and his colleagues.

4 And Beaverstock et al. (1999: 455) note in their roster of world cities that 'Africa has its first city in our list, Johannesburg, but there are still no world cities found in South Asia or the Middle East'.

5 It is perhaps appropriate here to raise some questions about the data used to classify world cities – data which, in Beaverstock et al. (1999) includes no Japanese banks and only US/UK/Australian and Canadian law firms. These firms may be the largest in the largest economies in the world, and thus arguably driving the global economy, but this methodology omits potentially significant dimensions of different economic globalisations and will fail to capture the regional significance of some centres, for example in the context of Asian and African cities (see Taylor 2001).

6 The example of Mumbai, where firms have a strong orientation in this data set to links with firms in non-core cities to the East, suggests that alternative sets of services firms might indicate alternative networks (Taylor 2004: 113).

7 It is interesting that the strongly national focus of advanced economies through the middle decades of the century was not matched by cities in poorer countries, which were closely linked to global flows in relations of dependency and even in relations of autarky, as import substitution often meant branch plants of major companies were forced to locate within national economies. The newness of globalisation for American cities may not be matched by the economic histories of other contexts.

8 See especially Abu-Lughod (1999) and acknowledged in Sassen (2001a: 28) and Taylor (2004: 58–9).

9 Of course there are many ideas about urban development that circulate through cities, and an ordinary city perspective would bring the engagements across these diverse narratives and strategies for shaping urban futures into view (Clarke and Gaile 1997).

10 As Saskia Sassen commented at a recent conference (Urban Futures, Johannesburg, 2000), she is often asked to advise city authorities on their development plans and always advises them to look to the specific advantages of that city, rather than be driven by an externally derived set of ambitions. And in many ways global- and world-cities approaches are rich critiques of the growth paths of cities: their use beyond the academy, though, has alighted on the hierarchical relations among cities and the globalising processes identified by theorists to propose intentional growth paths to match these accounts.

5 BRINGING THE CITY BACK IN: BEYOND DEVELOPMENTALISM AND GLOBALISATION

1 Although the data on this are not very robust, see Sassen 2001a, and Storper 1997.

2 See Scott et al. (2001: 23–6) and also Scott and Storper (2003).

3 As Hall and Pfeiffer (2000: 121) note for the south-east of England; see also Buck et al. (2002).

4 See for example, Wirth (1964), especially Chapters 11 and 14.

5 See for example, Balbo 1993, Allen et al. 1999, Graham and Marvin 2001, Amin and Thrift 2002, Le Galès 2002, Caldeira, 2000.

6 Although some have found creative ways to apply it, for example, Tyner 2000.

7 See Ferguson 1990, Escobar 1995; for recent reviews and examples see Pugh 1995, Gugler 1997, Burgess et al. 1997, Drakakis-Smith 2000.

8 These have been rehearsed regularly, but with little apparent effect on the wider literature: Simon 1989, King 1990, McGee 1995, Drakakis-Smith 2000.

9 Available online at <http://www.ukinindia.com/htdocs/dfid.asp#h>, accessed 23 May 2004; see also DfID 2001: 35.

10 For example, Logan and Molotch 1987, Cox and Mair 1988, Harvey 1989, Hall and Hubbard 1998.

11 See for example, McCarney 2000, World Bank 2000, Rogerson and Rogerson 1999, Rogerson 2000, Parnell and Pieterse 1998, Robinson 2002b.

12 For example, UNCHS 2001, World Bank 2000, DfID 2001.

13 Benjamin (2000) explores this dynamic for the Bangalore case.

14 Ketso Gordhan (former city manager of Johannesburg Metropolitan Council), interview, Johannesburg, 2 September 2002.

15 As Beall et al. (2002) note; Stan Thusini (former Northern Municipality councillor), interview, 20 July 2001.

16 This was later expanded to fifteen and became institutionalised within the municipality as the 'Transformation Lekgotla' – *lekgotla* is a vernacular term for a community gathering, or meeting.

17 Kenny Fihla (Mayoral Executive Committee), interview, 1 September 2002.

18 iGoli is the vernacular name for Johannesburg, meaning 'place of gold'.

19 Interview, 1 September 2002.

20 Interview, 2 September 2002.

21 Proceedings of the iGoli Summit, minutes of the iGoli Partnership Steering Committee, 3 September 1999.

22 Rashid Seedat (Project Leader, iGoli 2010), interview, 10 September 2002.

23 Prominently, efforts by a forum of inner-city officials, community leaders and business to revision the inner city, subject to considerable decline as businesses migrated to the northern parts of the city, produced a positive and upbeat image of Johannesburg as the 'Golden Heartbeat of Africa' (Bremner 2000). Other visionary initiatives drew on participatory local planning processes, mandated by national legislation and produced spatial development frameworks, which hoped to provide a framework for the expansion of business opportunities within a 'compact city' – using development corridors to channel economic activities through the city (GJMC 1998).

24 Lone Poulson (Community Technical Adviser, iGoli 2010), interview, 18 July 2001; Centre for the Development of the Built Environment (CDBE) 2000.

25 Stan Thusini (former councillor), interview, 20 July 2001.

26 City of Johannesburg 2001, minutes of stakeholder meetings, Rashid Seedat (Project Leader, iGoli 2010), interview, 10 June 2002.

27 The consultants who undertook the data review exercise for iGoli 2010 (at great expense to the city – 1 million US dollars) were a South African branch of Monitor Consultants, originally established by Michael Porter out of Harvard University, and now established in over twenty countries around the world. The City Manager had worked with them while director general in the Central Government's department of transport, and they had recently been appointed by the City of Durban in South Africa to undertake a more limited economic analysis of that city. They brought with them their standard arguments about the competitiveness of firms, drawing on a sectoral and cluster-based economic analysis using existing industrial census data and also coordinated an extensive review of the different areas of council responsibility (including health, property, waste, water and electricity provision).

28 Stan Thusini (former councillor), interview, 20 July 2001.

29 Marius de Jager (CEO Johannesburg Chamber of Commerce), interview, Johannesburg, 1 September 2001. The white ratepayers' representatives refused to be interviewed and were patently annoyed at the process having been terminated.

30 Personal communication, July 2001.
31 Rashid Seedat (Project Leader, iGoli 2010), interview, 2 September 2002; Sandy Lowitt (Consultant, author *Joburg 2030*), interview, 14 June 2002.
32 As compared to developmentalist approaches, which tend to concentrate on creating participatory forums to secure democratic and 'pro-poor' city visioning (Parnell and Robinson 2005).
33 Khetso Gordhan (former city manager), interview, 2 September 2002.
34 The new unicity has been subdivided into eleven administrative regions.
35 Sam Modiba (Administrator, Region 6), Interview 6 September 2002; Rashid Seedat (Project Leader, iGoli 2010), Interview, 10 September 2002.

6 CITY FUTURES: URBAN POLICY FOR ORDINARY CITIES

1 Kenny Fihla (Mayoral Executive Committee), interview, 1 August 2002. A slightly more cynical view came from Khetso Gordhan, former city manager, who suggested that most of the basic ideas were well known – people had written books about successful cases – and 'So we got a copy of the book, read it and we said, we know how to do this, we'll do it here. There were, there were definitely ideas, but we had picked up a lot of stuff already,' interview, 2 August 2002).
2 See for example Logan and Molotch 1987, Cox and Mair 1988, Brenner 2002.
3 See Buck et al. 2002 on London, Sites 2003 on New York, Kelly 2000 on Manila.
4 For example, Harloe 1981, DiGaetano and Klemanski 1993, Stoker and Mossberger 1994.
5 See Logan and Molotch 1987, Cox and Mair 1988, Jonas 1991.
6 See Boddy and Parkinson 2004.
7 With thirty-three local authorities in Greater London, ten in Manchester and three in Glasgow for example.
8 In Johannesburg, both of these issues have been shown to be crucial determinants of future investment levels by existing firms (Chandra et al. 2001).
9 This vision was developed in the context of a visit from officials and elected representatives from the city of Dayton, Ohio, and with technical support from South African-based consultants.
10 See Gordon and McCann (2000) for a review.
11 Schmitz and Nadvi (1999: 1504) note that 'developing country experiences are substantially different from the Italian model'.
12 This is because although only 38 per cent of firms were not involved in export markets, only 5 per cent had export markets as their major markets (Rogerson 2001: 43).
13 The survey was conducted as part of the iGoli planning initiative with an explicit ambition to consider whether the Council could play a more direct role in supporting employment generation in a situation of slow and relatively jobless growth (Foreword, in Chandra et al. 2001). Some of the core findings of this study informed the *Joburg 2030* policy statement discussed earlier in this chapter, including the assessment that crime was a key barrier to the expansion of firms. But more detailed evidence suggests that were they to expand, most firms across all sectors of the economy would choose to remain on their exact site, in the Johannesburg area or in the broader Gauteng region.
14 In the case of Bangalore, Benjamin (2000) has observed the existence of important substantive links between formal globally oriented firms and informal activities there, including recycling of computing equipment. He also reports at length on the enclaves of diverse kinds of activity that characterise established indigenous economies, as opposed to the high-tech transnational firms that create segregated business parks for their operations. In the indigenous enclaves, dense networks of interaction amongst quite different kinds of firms and social groups sustain a

dynamic economy. Although these kinds of activities might not be cutting edge and globally competitive in the sense of formal economic indicators, they expand the economy and offer opportunities for access to livelihoods and can involve exports both to a national hinterland and abroad.

CONCLUSION

1   As Amin and Thrift note, 'of course we write this with Northern cities in mind' (2002: 5); and James Donald notes, for example, that although Simmel offered no detailed observations of any city in his writings – that his was a rather 'unreal' city – 'We only know that Simmel has Berlin primarily in mind when he muses about what would happen if all its clocks stopped at once, although he is also happy to refer to London when it serves his turn' (1999: 10)
2   Some of the issues involved in this distinction are helpfully discussed by various authors in Cheah and Robbins 1998.
3   Scott and Storper (2003: 582) note that they 'seek to build a common theoretical language about the development of regions and countries in all parts of the world' while at the same time they want to recognise that 'territories are arrayed at different points along a vast spectrum of developmental characteristics'. This goes a long way to addressing my concerns with the categories and divisions of urban studies. And although it does preserve the suggestion that cities are on some trajectory of development, I would agree that it is also realistic to be alert to the very unequal circumstances within which urban life and urban development interventions take place.

Bibliography

Abers, R. (1998) 'Learning Democratic Practice: Distributing Government Resources through Popular Participation in Porto Alegre, Brazil', in M. Douglass and J. Friedmann (eds) *Cities for Citizens*, London: John Wiley, pp. 39–66.

Abers, R. (2004) 'Participatory Watershed Management in Brazil', paper presented to Workshop on Second Cities in a Global World, Cornell University, Department of City and Regional Planning, May.

Abu-Lughod, J. L. (1961) 'Migrant Adjustment to City Life: The Egyptian Case', *American Journal of Sociology* 57, 1: 22–32.

Abu-Lughod, J. L. (1995) 'Comparing Chicago, New York and Los Angeles: Testing Some World City Hypotheses', in P. Knox and P. Taylor (eds) *World Cities in a World-System*, Cambridge: Cambridge University Press, pp. 171–91.

Abu-Lughod, J. L. (1999) *New York, Chicago, Los Angeles: America's Global Cities*, London and Minneapolis, Minn.: University of Minnesota Press.

Acioly, C. (2001) 'Reviewing Urban Revitalisation Strategies in Rio de Janiero: From Urban Project to Urban Management Approaches', *Geoforum* 32: 509–20.

African Development Bank (2000) *African Development Report*, Oxford: Oxford University Press, for the African Development Bank.

Allen, J. (1999) 'Cities of Power and Influence: Settled Formations', in J. Allen, D. Massey and M. Pryke (eds) *Unsettling Cities*, London: Routledge, pp. 181–228.

Allen, J. (2002) 'Living on Thin Abstractions: More Power/Economic Knowledge', *Environment and Planning A* 34: 451–66.

Allen. J. (forthcoming) 'The Cultural Spaces of Siegfried Kracauer: The Many Surfaces of Berlin', *New Formations*.

Allen, J. and N. Henry (1995) 'Growth at the Margins: Contract Labour in a Core Region', in C. Hadjimichaelis and D. Sadler (eds) *Europe at the Margins*, London: Wiley, pp. 148–66.

Allen, J., D. Massey and M. Pryke (eds) (1999) *Unsettling Cities*, London: Routledge.

Altenberg, T. and J. Meyer-Stamer (1999) 'How to Promote Clusters: Policy Experiences from Latin America', *World Development* 27, 9: 1693–713.

Amin, A. (2003) 'Spaces of Corporate Learning', in J. Peck and H. Yeung (eds) *Remaking the Global Economy*, London: Sage, pp. 114–29.

Amin, A. and S. Graham (1997) 'The Ordinary City', *Transactions of the Institute of British Geographers* 22: 411–29.

Amin, A. and N. Thrift (1992) 'Neo-Marshallian Nodes in Global Networks', *International Journal of Urban and Regional Research* 4: 571–87.

Amin, N. and N. Thrift (2002) *Cities: Reimagining the Urban*, Cambridge: Polity.

Amis, P. (2002) 'Municipal Government, Urban Economic Growth and Poverty: Identifying the Transmission Mechanisms Between Growth and Poverty', in C. Rakodi and T. Lloyd-Jones (eds) *Urban Livelihoods: A People-Centred Approach to Reducing Poverty*, London: Earthscan, pp. 97–111.

Andrusz, G., M. Harloe and I. Szelenyi (eds) (1996) *Cities after Socialism*, Oxford: Blackwell.

Appadurai, A. (1996) *Modernity at Large: Cultural Dimensions of Globalization*, Minneapolis, Minn.: University of Minnesota Press.

Askew, M. and W. Logan (1994) *Cultural Identity and Urban Change in Southeast Asia: Interpretative Essays*, Victoria: Deakin University Press.

Balbo, M. (1993) 'Urban Planning and the Fragmented City of Developing Countries', *Third World Planning Review* 15: 23–35.

Beall, J., O. Crankshaw and S. Parnell (2002) *Uniting a Divided City: Governance and Social Exclusion in Johannesburg*, London: Earthscan.

Beall, J. (2000) 'Life in the Cities', in T. Allen and A. Thomas (eds) *Poverty and Development into the 21st Century*, Oxford: Oxford University Press, pp. 425–43.

Beauregard, R. and J. Pierre (2000) 'Disputing the Global: A Sceptical View of Locality-Based International Initiatives', *Policy and Politics* 28: 465–78.

Beaverstock, J., R. G. Smith and P. J. Taylor (2000) 'World-City Network: A New Metageography?' *Annals of the Association of American Geographers* 90: 123–34.

Beaverstock, J., P. Taylor and R. Smith (1999) 'A Roster of World Cities', *Cities* 16, 6: 444–58.

Beavon, K. (1998) ' "Johannesburg": Coming to Grips with Globalization from an Abnormal Base', in F. Lo and Y. Yeung (eds) *Globalization and the World of Large cities*, New York: United Nations Press, pp. 352–88.

Benjamin, S. (2000) 'Governance, Economic Settings and Poverty in Bangalore', *Environment and Urbanisation*, 12: 35–56.

Benjamin, W. (1979) *One Way Street*, London: Verso.

Benjamin, W. (1997) 'The Paris of the Second Empire in Baudelaire', in *Charles Baudelaire*, London: Verso, pp. 9–106.

Benjamin, W. (1999a) *The Arcades Project*, trans. H. Eiland and K. McLaughlin, Cambridge, Mass.: Harvard University Press.

Benjamin, W. (1999b) [1940] 'Theses on the Philosophy of History' in H. Arendt (ed.) *Illuminations*, London: Pimlico, pp. 245–55.

Benjamin, W. (1999c) 'On Some Motifs in Baudelaire', in H. Arendt (ed.) *Illuminations*, London: Pimlico, pp. 152–96.

Benton, T. (2003) 'Art Deco Architecture', in C. Benton, T. Benton and G. Wood (eds) *Art Deco, 1910–1939*, London: V & A Publications, pp. 245–59.

Berman, M. (1983) *All That is Solid Melts Into Air: The Experience of Modernity*, London: Verso.

Berner, E. and R. Korff (1995) 'Globalisation and Local Resistance: The Creation of Localities in Manila and Bangkok', *International Journal of Urban and Regional Research* 19: 208–22.

Boddy, M. and M. Parkinson (2004) *City Matters: Competitiveness, Cohesion and Urban Governance*, Bristol: Policy Press.

Boden, D. and H. Molotch (1993) 'The Compulsion of Proximity', in D. Boden and R. Friedland (eds) *Now/here: Time, Space and Social Theory*, Berkeley, Calif.: University of California Press, pp. 257–86.

Bond, P. (2003) *Against Global Apartheid*, London: Zed Press.

Bonick, G. C. (1997) *Zambia Country Assistance Review: Turning an Economy Around*, Washington, DC: World Bank.

Borges, J. (1964) 'The Fearful Sphere of Pascal', in *Labyrinths: Selected Stories and Other Writings*, edited by Donald A. Yates and James E. Irby, New York: New Directions, pp. 189–92.

Boyer, R. (2000) 'Is a Finance-Led Growth Regime a Viable Alternative to Fordism? A Preliminary Analysis', *Economy and Society* 29: 111–45.

Bremner, L. (2000) 'Reinventing the Johannesburg Inner City', *Cities* 17: 185–93.

Brenner, N. (1998) 'Global Cities, Glocal States: Global City Formation and State Territorial Restructuring in Contemporary Europe', *Review of International Political Economy*, 5: 1–37.

Brenner, N. (2002) 'Decoding the Newest "Metropolitan Regionalism" in the USA: A Critical Overview', *Cities* 19, 1: 3–21.

Browder, J. O. and B. J. Godfrey (1997) *Rainforest Cities: Urbanisation, Development and Globalization of the Brazilian Amazon*, New York: Columbia University Press.

Buck, N., I. Gordon, P. Hall, M. Harloe and M. Kleinman (2002) *Working Capital. Life and Labour in Contemporary London*, London: Routledge.

Buck-Morss, S. (1986) 'The Flaneur, the Sandwichman and the Whore: The Politics of Loitering', *New German Critique* 39: 99–141.

Buck-Morss, S. (1989) *The Dialectics of Seeing: Walter Benjamin and the Arcades Project*, Cambridge, Mass.: MIT Press.

Bunnell, T. (1999) 'Views from Above and Below: The Petronas Twin Towers and Contesting Visions of Development in Contemporary Malaysia', *Singapore Journal of Tropical Geography* 20: 1–23.

Bunnell, T. (2002) '*Kampung* Rules: Landscape and the Contested Government of Urban(e) Malayness', *Urban Studies* 39: 1685–701.

Bunnell, T. (2004a) *Malaysia, Modernity and the Multimedia Super Corridor: A Critical Geography of Intelligent Landscapes*, London and New York: Routledge Curzon.

Bunnell, T. (2004b) 'Re-Viewing the *Entrapment* Controversy: Megaprojection, (Mis)representation and Post-colonial Performance', *Geojournal* 59: 297–305.

Burgess, R., M. Carmona and T. Kolstee (eds) (1997) *The Challenge of Sustainable Cities: Neo-Liberalism and Urban Strategies in Developing Countries*, London: Zed Books.

Caldeira, T. (2000) *City of Walls: Crime, Segregation and Citizenship in São Paulo*, Berkeley, Calif.: University of California Press.

Campbell, T. (1997) *Innovations and Risk Taking: The Engine of Reform in Local Government in Latin America and the Caribbean*, World Bank Discussion Paper, No. 357, Washington, DC: World Bank.

Campbell, T. (1999) 'The Changing Prospects for Cities in Development: The Case of Vietnam', in *Business Briefing: World Urban Economic Development*, Official briefing for World Competitive Cities Congress, Washington, DC: World Bank, pp. 16–19.

Cardoso, F. (2003) 'Ambiguously Modern: Art Deco in Latin America', in C. Benton, T. Benton and G. Wood (eds) *Art Deco, 1910–1939*, London: V & A Publications, pp. 396–405.

Castells, M. (1977) *The Urban Question*, London: Edward Arnold.

Castells, M. (1983) *The City and the Grassroots*, London: Edward Arnold.

Castells, M. (1996) *The Rise of the Network Society*, Cambridge, Mass.: Blackwell.

Cawthorne, P. (1995) 'Of Networks and Markets: The Rise and Rise of a South Indian Town, the Example of Tiruppur's Cotton Knitwear Industry', *World Development*, 23: 43–56.

Caygill, H. (1998) *Walter Benjamin: The Colour of Experience*, London: Routledge.

Centre for the Development of the Built Environment (Johannesburg) (2000) *iGoli 2010: Visioning Process. Focus Group Workshops*, report available from CDBE, Faculty of the Built Environment, University of the Witwatersrand.

Chakrabarty, D. (2000) *Provincialising Europe*, London: Routledge.

Chandra, V., L. Moorty, B. Rajaratnam and K. Schaefer (2001) *Constraints to Growth and Employment in South Africa. Report No. 1: Statistics from the Large Manufacturing Firm Survey* Washington, DC: World Bank, Informal Discussion Papers on Aspects of the Economy of South Africa.

Chandra, V., J.-P. Nganou and C. M. Noel (2002) *South Africa: Constraints to Growth in Johannesburg's Black Informal Sector*, Washington, DC: World Bank, Informal Discussion Papers on Aspects of the Economy of South Africa.

Cheah, P. and B. Robbins (eds) (1998) *Cosmopolitics: Thinking and Feeling Beyond the Nation*, Minneapolis, Minn.: University of Minnesota Press.

City of Johannesburg (2001) *Johannesburg: An African City in Change*, Cape Town: Zebra Press.

Cities Alliance (2002) 'City Development Strategies: First Result', available on-line at <http://www.citiesalliance.org>. Accessed 15 July 2004.

Clark, J., with C. Allison (1989) *Zambia. Debt and Poverty*, Oxford: Oxfam.

Clarke, S. E. and G. L. Gaile (1997) 'Local Politics in a Global Era: Thinking Locally, Acting Globally', *Annals of the American Academy of Political and Social Science* 551: 28–43.

Clifford, J. (1989) '1933, February. Negrophilia', in D. Hollier (ed.) *A New History of French Literature*, Cambridge, Mass.: Harvard University Press, pp. 901–8.

Clifford, J. (1997) *Routes: Travel and Translation in the late Twentieth Century*, Cambridge, Mass.: Harvard University Press.

Cohen, M. A. (1997) 'The Hypothesis of Urban Convergence: Are Cities in the North and South Becoming More Alike in an Age of Globalisation?' in M. A. Cohen, B. A. Ruble, J. S. Tulchin and A. M. Garland (eds) *Preparing for the Urban Future: Global Pressures and Local Forces*, Washington, DC: Woodrow Wilson Centre Press.

Comaroff, J. and J. Comaroff (1997) *Modernity and its Malcontents*, Chicago, Ill.: Chicago University Press.

Cowen, M. and R. Shenton (1996) *Doctrines of Development*, London: Routledge.

Cowley, M. (1994) *Exiles Return*, London: Penguin.

Cox, K. and A. Mair (1988) 'Locality and Community in the Politics of Local Economic Development', *Annals, Association of American Geographers* 78: 307–25.

Crysler, G. (2003) *Writing Spaces: Discourses in Architecture, Urbanism and the Built Environment, 1960–2000*, London: Routledge.

de Certeau, M. (1984) *The Practice of Everyday Life*, Berkeley, Calif.: University of California Press.

Devons, E. and M. Gluckman (1964) 'Introduction', in M. Gluckman (ed.) *Closed Systems and Open Minds: The Limits of Naivety in Social Anthropology*, Edinburgh and London: Oliver & Boy, pp. 13–19.

Dewey, J. (1960) 'The Rural-Urban Continuum: Real but Relatively Unimportant', *American Journal of Sociology* 66, 1: 60–6.

DfID (Department for International Development) (2001) *Meeting the Challenge of Poverty in Urban Areas*, London: DfID.

Dhlomo, H. I. E. (1985) *Collected Works*, edited by N. Visser and T. Couzens, Johannesburg: Ravan Press.

Dick, H. W. and P. J. Rimmer (1998) 'Beyond the Third World City: The New Urban Geography of Southeast Asia', *Urban Studies* 35: 2303–21.

DiGaetano, A. and J. Klemanski (1993) 'Urban Regimes in Comparative Perspective: The Politics of Urban Development in Britain', *Urban Affairs Quarterly* 29: 54–83.

Domosh, M. (1990) 'Those "Sudden Peaks That Scrape the Sky": The Changing Imagery of New York's First Skyscrapers', in L. Zonn (ed.) *Place Images in Media: Portrayal, Experience, and Meaning*, Savage, Md.: Rowan & Littlefield, pp. 9–29.

Domosh, M. (1992) 'Corporate Cultures and the Modern Landscape of New York City', in K. Anderson and F. Gale (eds) *Inventing Places: Studies in Cultural Geography*, London: Longman, pp. 72–86.

Donald, J. (1999) *Imagining the Modern City*, London: The Athlone Press.

Douglas, A. (1996) *Terrible Honesty: Mongrel Manhattan in the 1920s*, London: Picador.

Douglass, M. (1998) 'World city formation in the Asia Pacific Rim: Poverty, "Everyday" Forms of Civil Society and Environmental Management', in M. Douglass and J. Friedmann (eds) *Cities for Citizens*, Chichester: John Wiley, pp. 107–37.

Douglass, M. (2000) 'The Rise and Fall of World Cities in the Changing Space-Economy of Globalisation: Comment on Peter J. Taylor's "World Cities and Territorial States under Conditions of Contemporary Globalisation" ', *Political Geography* 19: 43–9.

Douglass, M. (2001) 'Inter-City Competition and the Question of Economic Resilience: Globalization and the Asian Crisis', in A. J. Scott (ed.) *Global City-Regions*, Oxford: Oxford University Press, pp. 236–62.

Drakakis-Smith, D. (2000) *Third World Cities*, London: Routledge.

Duranton, G. and D. Puga (2000) 'Diversity and Specialisation in Cities: Why, Where and When Does it Matter?' *Urban Studies* 37: 533–55.

Duranton, G. and D. Puga (2001) 'Nursery Cities: Urban Diversity, Process Innovation, and the Life Cycle of Products', *American Economic Review* 91: 1454–77.

Epstein, A. L. (1958) *Politics in an Urban African Community*, Manchester: Manchester University Press and Institute for Social Research, Lusaka, Zambia.

Epstein, A. L. (1964) 'Urban Communities in Africa', in M. Gluckman (ed.) *Closed Systems and Open Minds: The Limits of Naivety in Social Anthropology*, Edinburgh and London: Oliver & Boy, pp. 83–102.

Epstein, A. L. (1969) 'The Network and Urban Social Organisation', in J. C. Mitchell (ed.) *Social Networks in Urban Situations: Analyses of Personal Relationships in Central African Towns*, Manchester: Manchester University Press and Institute for Social Research, University of Zambia, Lusaka, pp. 77–116.

Escobar, A. (1995) *Encountering Development: The Making and Unmaking of the Third World*, Princeton, NJ: Princeton University Press.

Etzioni, A. (1959) 'The Ghetto: A Re-Evaluation', *Social Forces* 37: 255–62.

Evers, H.-D. and R. Korff (2000) *Southeast Asian Urbanism: The Meaning and Power of Social Space*, Münster: Lit. Verlag and Singapore: Institute of Southeast Asian Studies.

Fabian, J. (1983) *Time and the Other*, New York: Columbia University Press.

Fainstein, S., I. Gordon and M. Harloe (eds) (1992) *Divided Cities*, Oxford: Blackwell.

Federal Writers' Project (1939) *New York Panorama: A Comprehensive View of the Metropolis*, London: Constable.

Felski, R. (1995) *The Gender of Modernity*, London: Routledge.

Ferguson, J. (1990) *The Anti-Politics Machine*, Minneapolis, Minn.: University of Minnesota Press.

Ferguson, J. (1999) *Expectations of Modernity: Myths and Meanings of Urban Life on the Zambian Copperbelt*, Berkeley, Calif.: University of California Press.

Firman, T. (1999) 'From "Global City" to "City of Crisis": Jakarta Metropolitan Region Under Economic Turmoil', *Habitat International* 23: 447–66.

Fraser, V. (2000) *Building the New World: Studies in the Modern Architecture of Latin America, 1930–1960*, London: Verso.

Friedmann, J. (1995a) 'The World City Hypothesis', in P. Knox and P. Taylor (eds) *World Cities in a World System*, Cambridge: Cambridge University Press, pp. 317–31. Originally published in 1986 in *Development and Change* 17: 69–84.

Friedmann, J. (1995b) 'Where We Stand Now: A Decade of World City Research', in P. Knox and P. Taylor (eds) *World Cities in a World System*, Cambridge: Cambridge University Press, pp. 21–48.

Friedmann, J. and W. Goetz (1982) 'World City Formation. An Agenda for Research and Action', *International Journal of Urban and Regional Research* 6: 309–44.

Friedmann, J. and R. Wulff (1976) *The Urban Transition: Comparative Studies of Newly Industrialising Societies*, London: Edward Arnold.

Frisby, D. (2001) *Cityscapes of Modernity*, Cambridge: Polity.

Gans, H. (1962) *The Urban Villagers: Group and Class in the Life of Italian-Americans*, New York: The Free Press.

Gans, H. (1995) 'Urbanism and Suburbanism as Ways of Life: A Reevaluation of Definitions', in P. Kasinitz (ed.) *Metropolis: Center and Symbol of Our Times*, New York: New York University Press.

GHK Consultants (2000) *City Development Strategies: Taking Stock and Signposting the Way Forward*, A Discussion Report for Department for International Development (UK) and the World Bank, London.

GHK Consultants (2002) *City Development Strategies: An Instrument for Poverty Reduction?* Final Report to Department for International Development (UK). August.

Gilbert, A. (1998) 'World Cities and the Urban Future: The View from Latin America', in F. Lo and Y. Yeung (eds) *Globalisation and the World of Large Cities*, Tokyo: UN University Press, chapter 8.

Gilroy, P. (1993) *The Black Atlantic*, London: Verso.

Gluckman, M. (1961) 'Anthropological Problems Arising from the African Industrial Revolution', in A. Southall (ed.) *Social Change in Modern Africa*, London, New York and Toronto: Oxford University Press, for International African Institute, Kampala, pp. 67–82.

Goankar, D. (2001) 'On Alternative Modernities', in D. Goankar (ed.) *Alternative Modernities*, Durham, NC and London: Duke University Press, pp. 1–23.

Goh, Beng Lan (2002) *Modern Dreams: An Inquiry into Power, Cultural Production and the Cityscape in Contemporary Urban Penang, Malaysia*, Ithaca, NY: Cornell South East Asia Program Publications.

Goldfrank, B. (2003) 'Making Participation Work in Porto Alegre', in G. Baiocchi (ed.) *Radicals in Power: The Workers' Party (PT) and Experiments in Urban Democracy in Brazil*, London: Zed Books.

Gordon, I. (1999) 'Internationalisation and Urban Competition', *Urban Studies* 36: 1001–16.

Gordon, I. (2003) 'Capital Needs, Capital Growth and Global City Rhetoric in Mayor Livingstone's London Plan', paper presented at the Annual Conference of American Geographers, New Orleans, March.

Gordon, I. and P. McCann (2000) 'Industrial Clusters: Complexes, Agglomeration and/or Social Networks', *Urban Studies*, 37: 513–32.

Gotz, G. and A. Simone (2003) 'On Belonging and Becoming in African Cities', in R. Tomlinson, R. Beauregard, L. Bremner and X. Mangcu (eds) *Emerging Johannesburg*, London: Routledge, pp. 123–47.

Graham, S. and S. Marvin (2001) *Splintering Urbanism: Networked Infrastructures, Technological Mobilities and the Urban Condition*, London and New York: Routledge.

GJMC (Greater Johannesburg Metropolitan Council) (1998) *Spatial Development Framework*, Johannesburg.

GJMC (Greater Johannesburg Metropolitan Council) (2000) *Igoli 2002: Making the City Work*, Johannesburg.

GJMC (Greater Johannesburg Metropolitan Council) (2002) *Joburg 2030*, Johannesburg.

Gugler, J. (ed.) (1997) *Cities in the Developing World: Issues, Theory and Policy*, Oxford: Oxford University Press.

Gupta, A. and J. Ferguson (eds) (1999) *Culture, Power, Place: Explorations in Cultural Anthropology*, Durham, NC and London: Duke University Press.

Halfani, M. (1996) 'Marginality and Dynamism: Prospects for the Sub-Saharan African City', in M. A. Cohen, B. A. Ruble, J. S. Tulchin and A. M. Garland (eds) *Preparing for the Urban Future: Global Pressures and Local Forces*, Washington, DC: Woodrow Wilson Centre Press, chapter 5.

Hall, P. (1966) *The World Cities*, London: Weidenfeld & Nicholson.

Hall, P. (2001) 'Global City-Regions in the Twenty-First Century', in A. Scott (ed.) *Global City-Regions: Trends, Theory, Policy*, Oxford: Oxford University Press, pp. 59–77.

Hall, P. and U. Pfeiffer (2000) *Urban Future 21: A Global Agenda for Twenty-First Century Cities*, London: E&FN Spon.

Hall, T. and P. Hubbard (eds) (1998) *The Entrepreneurial City: Geographies of Politics, Regime and Representation*, Chichester, Wiley.

Hannerz, U. (1980) *Exploring the City: Inquiries Towards an Urban Anthropology*, New York: Columbia University Press.

Hansen, K. (1994) 'Dealing with Used Clothing: *Salaula* and the Construction of Identity in Zambia's Third Republic', *Public Culture* 6: 503–23.

Hansen, K. (1997) *Keeping House in Lusaka*, New York: Columbia University Press.

Hansen, K. (2000) *Salaula: The World of Secondhand Clothing and Zambia*, Chicago, Ill.: Chicago University Press.

Harding, A., I. Deas, R. Evans and S. Wilks-Heeg (2004) 'Reinventing Cities in a Restructuring Region? The Rhetoric and Reality of Renaissance in Liverpool and Manchester', in M. Boddy and M. Parkinson (eds) *City Matters: Competitiveness, Cohesion and Urban Governance*, Bristol: Policy Press, pp. 33–50.

Harloe, M. (1981) 'Notes on Comparative Urban Research', in M. Dear and A. Scott (eds) *Urbanisation and Urban Planning in Capitalist Society*, London: Methuen.

Harries-Jones, P. (1975) *Freedom and Labour: Mobilization and Political Control on the Zambian Copperbelt*, Oxford: Basil Blackwell.

Harris, N. (1986) *The End of the Third World*, Harmondsworth: Penguin.

Harris, N. (ed.) (1992) *Cities in the 1990s*, London: UCL Press.

Harris, N. (1995) 'Bombay in a Global Economy: Structural Adjustment and the Role of Cities', *Cities* 12: 175–84.

Harris, N. (2002) 'Cities as Economic Development Tools', *Urban Brief*, Washington, DC: Woodrow Wilson International Centre for Scholars.

Harvey, D. (1989) 'From Managerialism to Entrepreneurialism: The Transformation of Urban Governance in Late Capitalism', *Geografiska Annaler* 71B: 3–17.

Hausner, P. M. (1965) Observations on the urban-folk and urban-rural dichotomies as forms of Western ethnocentrism, in P. M. Hausner and L. F. Schnore (eds) *The Study of Urbanisation*, London: John Wiley and Sons, pgs 503–517.

Hausner, P. M. and L. F. Schnore (eds) (1965) *The Study of Urbanisation*, London: John Wiley and Sons.

Healey, P., A. Khakee, A. Motte and B. Needham (eds) (1997) *Making Strategic Spatial Plans: Innovation in Europe*, London: University College London Press.

Henderson, V. (2002) 'Urbanization in Developing Countries', *The World Bank Research Observer* 17: 89–112.

Hertz, E. (2001) 'Face in the Crowd: The Cultural Construction of Anonymity in Urban China', in N. Chen, C. D. Clark, S. Gottschang and L. Jeffery (eds) *China Urban: Ethnographies of Contemporary Culture*, Durham, NC and London: Duke University Press, pp. 274–94.

Hewitt, T. (2000) 'Half a Century of Development', in *Poverty and Development into the 21st Century*, Oxford: Oxford University Press, pp. 289–309.

Hill, R. C. and J. W. Kim (2000) 'Global Cities and Developmental States: New York, Tokyo and Seoul', *Urban Studies* 37: 2167–95.

Holleran, M. (1996) 'Boston's "Sacred Skyline": From Prohibiting to Sculpting Skyscrapers, 1891–1928', *Journal of Urban History*, 27: 552–85.

Jacobs, J. (1965) [1961] *The Death and Life of Great American Cities*, Harmondsworth: Penguin.

Jacobs, J. (1972) [1969] *The Economy of Cities*, Harmondsworth: Penguin.

Jacobs, J. (1996) *Edge of Empire: Postcolonialism and the City*, London: Routledge.

Jaguaribe, J. (1999) 'Modernist Ruins: National Narratives and Architectural Forms', *Public Culture* 11: 294–312.

James, D. (1999) *Songs of the Women Migrants: Performance and Identity in South Africa*, London: Edinburgh University Press for the International African Institute.

Jeshudasan, J. V. (1995) 'Statist Democracy and the Limits to Civil Society in Malaysia', *Journal of Commonwealth and Comparative Politics*, 33: 335–56.

Jessop, B. (1994) 'Post-Fordism and the State', in A. Amin (ed.) *Post-Fordism: A Reader*, Oxford: Basil Blackwell, pp. 251–79.

Jessop, B. and N. Sum (2000) 'An Entrepreneurial City in Action: Hong Kong's Emerging Strategies in and for (Inter) Urban Competition', *Urban Studies* 37: 2287–313.

Johnson, R. (1999) 'Brazilian Modernism: An Idea Out of Place?', in A. L. Geist and

J. B. Morleón (eds) *Modernism and its Margins: Reinscribing Cultural Modernity from Spain and Latin America*, London: Garland Publishing, pp. 186–214.

Jonas, A. (1991) 'Urban Growth Coalitions and Urban Development Policy: Postwar Growth and the Politics of Annexation in Metropolitan Columbus', *Urban Geography*, 12: 197–225.

Jonas, A. and K. Ward (2002) 'A World of Regionalisms? Towards a US-UK Urban and Regional Policy Framework Comparison', *Journal of Urban Affairs* 24, 4: 377–401.

Jones, A. (1998) 'Re-Theorising the Core: A 'Globalized' Business Elite in Santiago, Chile', *Political Geography* 17: 295–318.

Keil, R. (2000) 'Governance Restructuring in Los Angeles and Toronto: Amalgamation or Secession?', *International Journal of Urban and Regional Research* 24: 758–81.

Kelly, P. F. (2000) *Landscapes of Globalization: Human Geographies of Economic Change in the Philippines*, London: Routledge.

King, A. (1990) *Urbanism, Colonialism and the World-Economy*, London: Routledge.

King, A. (1995) Re-Presenting World Cities: Cultural Theory/Social Practice', in P. Knox and P. Taylor (eds) *World Cities in a World-System*, London: Routledge, pp. 215–31.

King, A. (1996) 'Worlds in the City: Manhattan Transfer and the Ascendance of Spectacular Space', *Planning Perspectives* 11: 97–114.

King, A. (2004) *Spaces of Global Cultures: Architecture Urbanism Identity*, London: Routledge.

Kirkpatrick, G. (2000) 'The Aesthetics of the Avant-Garde', in V. Schelling (ed.) *Through the Kaleidoscope: The Experience of Modernity in Latin America*, London: Verso, pp. 177–98.

Knox, P. (1995) 'World Cities in a World-System', in P. Knox and P. Taylor (eds) *World Cities in a World-System*. London: Routledge, pp. 3–20.

Knox, P. and P. Taylor (eds) (1995) *World Cities in a World-System*, London: Routledge.

Koolhaas, R. Harvard Project on the City, S. Boeri, S. Kwinter, N. Tazi and H. U. Obrist (eds) (2000) *Mutations*, Bordeaux: Actar.

Kraniauskas, J. (2000) 'Beware Mexican Ruins! "One Way Street" and the Colonial Unconscious', in A. Benjamin and P. Osborne (eds) *Walter Benjamin's Philosophy: Destruction and Experience*, Manchester: Clinamen Press, pp. 137–52.

Kuper, H. (ed.) (1965) *Urbanization and Migration in West Africa*, Berkeley, Calif.: University of California Press.

Kusno, A. (2000) *Behind the Post-colonial: Architecture, Urban Space and Political Cultures in Indonesia*, London: Routledge.

Laclau, E. (1990) *New Reflections on the Revolution of our Time*, London: Verso.

Lawson, V. and T. Klak (1993) 'An Argument for Critical and Comparative Research on the Urban Economic Geography of the Americas', *Environment and Planning A* 25: 1071–84.

Lee, L. O. (2001) 'Shanghai Modern: Reflections on Urban Culture in China in the 1930s', in D. Goankar (ed.) *Alternative Modernities*, Durham, NC and London: Duke University Press, pp. 86–122.

Le Galès, P. (2002) *European Cities: Social Conflicts and Governance*, Oxford: Oxford University Press.

Lewis, O. (1959) *Five Families*, New York: Basic Books.

Lewis, O. (1973) 'Some Perspectives on Urbanization with Special Reference to Mexico City', in A. Southall (ed.) *Urban Anthropology: Cross-Cultural Studies of Urbanization*, New York: Oxford University Press, pp. 125–38.

Lipietz, B. (2005) 'On the Openness of City Futures: Bringing Politics Back In', unpublished manuscript, Development Studies Centre, SOAS, London.

Li Puma, B. (2001) *Encompassing Others: The Magic of Modernity in Melanesia*, Ann Arbor, Mich.: University of Michigan Press.

Lloyd, P. C. (1973) 'The Yoruba: An Urban People?', in A. Southall (ed.) *Urban Anthropology: Cross-Cultural Studies of Urbanization*, New York: Oxford University Press, pp. 107–124.

Lo, F.-C. and Y.-M. Yeung (eds) (1998) *Globalisation and the World of Large Cities*, Tokyo: UN University Press.

Logan, J. and H. Molotch (1987) *Urban Fortunes: The Political Economy of Place*, Berkeley, Calif.: University of California Press.

London Development Agency (LDA) (2000) *London Development Strategy*, London: LDA.

Loughhead, S. and C. Rakodi (2002) 'Reducing Urban Poverty in India: Lessons from Projects Supported by DfID', in C. Rakodi with T. Lloyd-Jones (eds) *Urban Livelihoods: A People-Centred Approach to Reducing Poverty*, London: Earthscan, pp. 225–36.

Lusaka City Council (1999) *Five Year Strategic Plan*, Lusaka: Lusaka City Council.

Mabin, A. (1999) 'The Urban World through a South African Prism', in R. Beauregard and S. Body-Gendrot (eds) *The Urban Moment*, London: Sage, Urban Affairs Annual Reviews, pp. 141–52.

Machimura, T. (1998) 'Symbolic Use of Globalisation in Urban Politics in Tokyo', *International Journal of Urban and Regional Research* 30: 183–94.

Magubane, B. (1968) 'Crisis in African Sociology', *East African Journal*, December, pp. 21–40.

Malmberg, A. (2003) 'Beyond the Cluster: Local Milieus and Global Connections', in J. Peck and H. Yeung (eds) *Remaking the Global Economy*, London: Sage, pp. 145–62.

Marcuse, P. and R. van Kempen (2000) *Globalising Cities: A New Spatial Order?* Oxford: Blackwell.

Markusen, A. (1996a) 'Interaction between Regional and Industrial Policies: Evidence from Four Countries', *International Regional Science Review*, 19: 49–77.

Markusen, A. (1996b) 'Sticky Places in Slippery Space: A Typology of Industrial Districts', *Economic Geography*, 72: 293–313.

Markusen, A. and V. Gwiasda (1994) 'Multipolarity and the Layering of Functions in World Cities: New York City's Struggle to Stay on Top', *International Journal of Urban and Regional Research*, 18: 167–93.

Massey, D., J. Allen and S. Pile (eds) (1999) *City Worlds*, London: Routledge and the Open University Press.

Mayer, P. with contributions by I. Mayer (1971) [1961] *Townsmen or Tribesmen: Conservativism and the Process of Urbanization in a South African City*, Cape Town: Oxford University Press.

Mbembe, A. (2001) *On the Postcolony*, Berkeley, Calif.: University of California Press.

McCarney, P. (2000) *Disjunctures, Divides and Disconnects: The Promise of Local Government in Development*, Isandla Dark Roast Occasional Paper Series, Vol. II (1) September.

McClintock, A. (1995) *Imperial Leather*, London: Routledge.

McCormick, D. (1999) 'African Enterprise Clusters and Industrialization: Theory and Reality', *World Development* 27: 1531–51.

McGee, T. (1971) *The Urbanisation Process in the Third World: Explorations in Search of a Theory*, London: Bell.

McGee, T. (1995) 'Eurocentrism and Geography: Reflections on Asian Urbanisation', in J. Crush (ed.) *Power of Development*, London: Routledge.

Meade, T. A. (1997) *'Civilizing' Rio de Janeiro: Reform and Resistance in a Brazilian City, 1889–1930*, University Park, Pa.: Pennsylvania State University Press.

Miner, H. (1953) *The Primitive City of Timbuktoo*, Princeton, NJ: Princeton University Press.

Mitchell, J. C. (1956) *The Kalela Dance*, Rhodes Livingstone Papers No. 27. Lusaka: Rhodes-Livingstone Institute and Manchester University Press.

Mitchell, J. C. (1968) 'Theoretical Orientations in African Urban Studies', in M. Banton (ed.) *The Social Anthropology of Complex Societies*, London: Tavistock, pp. 37–68.

Mitchell, J. C. (ed.) (1969) *Social Networks in Urban Situations: Analyses of Personal Relationships in Central African Towns*, Manchester: Manchester University Press (with Institute for Social Research, University of Zambia).

Mitchell, J. C. (1973) 'Distance, Transportation and Urban Involvement in Zambia', in A. Southall (ed.) *Urban Anthropology*, New York: Oxford University Press, pp. 287–314.

Mitchell, J. C. (1987) *Cities, Society, and Social Perception: A Central African Perspective*, Oxford: Clarendon Press.

Mitchell, T. (ed.) (2000) *Questions of Modernity*, London and Minneapolis, Minn.: University of Minnesota Press.

Mitchell, T. and L. Abu-Lughod (1993) 'Questions of Modernity', *Items* 47: 79–83.

Monitor Group (2000) *iGoli 2010: Recommended Vision and Strategic Agenda*, Information Package for the Executive Committee of the Council of the City of Johannesburg, December.

Monitor Group (2001) *Towards a Strategy for Building Johannesburg into a World Class City: Proposed Strategic Framework for Development Through Delivery, Empowerment and Growth*, Discussion Document Commissioned by the City of Johannesburg, February.

Monitor Group (2002) *Durban at the Crossroads*, Durban Metropolitan Council.

Montgomery, M., R. Stren, B. Cohen, and H. E. Reed (2004) *Cities Transformed: Demographic Change and its Implications in the Developing World*, London: Earthscan.

Morris, A. (1999) *Bleakness and Light: Inner-City Transition in Hillbrow, Johannesburg*, Johannesburg: Witwatersrand University Press.

Morshidi, S. (2000) 'Globalising Kuala Lumpur and the Strategic Role of the Producer Services Sector', *Urban Studies*, 37: 2217–40.

Morton, P. A. (2000) *Hybrid Modernities: Architecture and Representation at the 1931 Colonial Exposition, Paris*, Cambridge, Mass.: MIT Press.

Moser, C. (1998) 'The Asset Vulnerability Framework: Reassessing Urban Poverty Reduction Strategies', *World Development* 26: 1–19.

Moser, C. and J. Holland (1997) *Household Responses to Poverty and Vulnerability. Vol IV. Confronting Crisis in Chawama, Lusaka, Zambia*, Washington, DC: World Bank for Urban Management Programme.

Mpe, P. (2001) *Welcome To Our Hillbrow*, Pietermaritzburg: University of Natal Press.

Mumford, L. (ed.) (1972) *Roots of Contemporary American Architecture*, New York: Dover.

Myers, G. (2003) *Verandahs of Power: Colonialism and Space in Urban Africa*, Syracuse, NY: Syracuse University Press.

Myers, G. (forthcoming) *Governance and Garbage: African Cities and Post-colonial Cultural Geography*, London: Ashgate Press.

Needell, J. D. (1987) *A Tropical Belle Epoque: Elite Culture and Society in Turn-of-the-Century Rio de Janeiro*, Cambridge: Cambridge University Press.

Ngwane, T. (2003) 'Sparks in the Township', *New Left Review* 22: 36–56.

Nye, D. E. (1992) 'Ritual Tomorrows: The New York World's Fair of 1939', *History and Anthropology* 6: 1–21.

Ogborn, M. (1998) *Spaces of Modernity: London's Geographies, 1680–1780*, New York: Guilford Press.

Olds, K. (1995) 'Globalization and the Production of New Urban Spaces: Pacific Rim Megaprojects in the Late 20th Century', *Environment and Planning A* 27: 1713–43.

Olds, K. and H. Yeung (2004) 'Pathways to Global City Formation: Views from the Developmental City-State of Singapore', *Review of International Political Economy* 11: 489–521.

Oliven, R. G. (2000) 'Brazil: The Modern in the Tropics', in V. Schelling (ed.) *Through the Kaleidoscope: The Experience of Modernity in Latin America*, London: Verso, pp. 53–74.

Oliviera, N. D. S. (1996) 'Favelas and Ghettos: Race and Class in Rio de Janeiro and New York City', *Latin American Perspectives* 91, 23: 71–89.

Ortiz, R. (2000) 'Popular Culture, Modernity and Nation', in V. Schelling (ed.) *Through the Kaleidoscope: The Experience of Modernity in Latin America*, London: Verso, pp. 127–47.

Osborne, P. (1992) 'Modernity is a Qualitative, not Chronological Category: Notes on the Dialectics of Different Historical Time', in F. Barker, P. Hulme and M. Iversen (eds) *Postmodernism and the Re-reading of Modernity*, Manchester: Manchester University Press, pp. 23–45.

Pahl, R. (1968) *Readings in Urban Sociology*, London: Pergamon Press.

Pahl, R. (1970) *Whose City? And Other Essays on Sociology and Planning*, London: Longman.

Park, R. E. (1952) [1914] *Human Communities: The City and Human Ecology*, New York: The Free Press.

Park, R. E. (1967) [1925] *On Social Control and Collective Behaviour*, edited and with an introduction by R. H. Turner, Chicago, Ill.: University of Chicago Press.

Parnell, S. (2002) *Johannesburg's City Development Strategy*, Unpublished Report for GHK Consultants, London.

Parnell, S. and E. Pieterse (1998) 'Developmental Local Government: The Second Wave of Post-Apartheid Reconstruction', *Africanus*, 29: 61–85.

Parnell, S. and J. Robinson (2005) 'Deciphering "Development" at the Urban Scale: Leads from Johannesburg's City Development Strategy', forthcoming, *Urban Studies*.

Pederson, P. O. and D. McCormick (1999) 'African Business Systems in a Globalising World', *Journal of Modern African Studies* 37: 109–135.

Pickvance, C. and E. Preteceille (eds) (1991) *State Restructuring and Local Power: A Comparative Perspective*, London: Pinter.

Piermay, J.-L. (1997) 'Kinshasa: A Reprieved Mega-City?' in C. Rakodi (ed.) *The Urban Challenge in Africa*, Tokyo: UN University Press, chapter 7.

Pile, S. (2005) *Real Cities: Modernity, Space and the Phantasmagorias of City Life*, London: Sage.

Pile, S., C. Brook and C. Mooney (eds) (1999) *Unruly Cities?* London: Routledge.

Pile, S. and N. Thrift (eds) (2000) *City A-Z*, London: Routledge.

Pillay, U. (2004) 'Are Globally Competitive "City Regions" Developing in South Africa? Formulaic Aspirations or New Imaginations?', *Urban Forum*, 15: 340–64.

Porter, M. (1996) 'Competitive Advantage, Agglomeration Economies, and Regional Policy', *International Regional Science Review*, 19: 85–94.

Porter, M., Monitor Group, on the Frontier and Council on Competitiveness (2002) *Clusters of Innovation: Regional Foundations of U.S. Competitiveness*, Washington, DC: Council on Competitivess.

Powdermaker, H. (1966) *Stranger and Friend: The Way of an Anthropologist*, New York: W.W. Norton.

Pryke, M. (1999) 'Open Futures', in J. Allen, D. Massey and M. Pryke (eds) *Unsettling Cities*, London: Routledge, chapter 8.

Pugh, C. (1995) 'Urbanisation in Developing Countries: An Overview of the Economic and Policy Issues in the 1990s', *Cities* 12: 381–98.

Rabinow, P. (1989) *French Modern*, Cambridge, Mass.: MIT Press.

Raco, M. (1999) 'Competition, Collaboration and the New Industrial Districts: Examining the Institutional Turn in Local Economic Development', *Urban Studies* 36: 951–68.

Rakodi, C. (ed.) (1997) *The Urban Challenge in Africa: Growth and Management of its Large Cities*, Tokyo: UN University Press.

Rakodi, C. with T. Lloyd-Jones (2002) *Urban Livelihoods: A People-Centred Approach to Reducing Poverty*, London: Earthscan.

Redfield, R. and M. Singer (1954) 'The Cultural Role of Cities', *Economic Development and Cultural Change* 3: 53–73.

Resende, B. (2000) 'Brazilian Modernism: The Canonised Revolution', in V. Schelling (ed.) *Through the Kaleidoscope: The Experience of Modernity in Latin America*, London: Verso, pp. 199–218.

Roberts, B. (1978) *Cities of Peasants*, London: Edward Arnold.

Robins, K. and A. Askoy (1996) 'Istanbul between Civilisation and Discontent', *City* 5–6: 6–33.

Robinson, C. and R. H. Bletter (1975) *Skyscraper Style: Art Deco New York*, New York: Oxford University Press.

Robinson, J. (1998) 'Spaces of Democracy: Re-mapping the Apartheid City', *Environment and Planning D: Society and Space* 16: 433–78.

Robinson, J. (2002a) 'Global and World Cities: A View from Off the Map', *International Journal of Urban and Regional Research*, 26: 531–54.

Robinson, J. (2002b) 'City Futures: New Territories for Development Studies?', in J. Robinson (ed.) *Development and Displacement*, Oxford: Oxford University Press.

Robinson, J. (2003a) 'Johannesburg's Futures: Between Developmentalism and Global Success', in R. Tomlinson, R. Beauregard, L. Bremner and X. Mangcu (eds) *Emerging Johannesburg*, London: Routledge, pp. 259–80.

Robinson, J. (2003b) 'Postcolonialising Geography: Tactics and Pitfalls', *Singapore Journal of Tropical Geography*, 24: 273–89.

Rodriguez-Posé, A., J. Tomaney and J. Klink (2001) 'Local Empowerment through Economic Restructuring in Brazil: The Case of the Greater ABC Region', *Geoforum* 32: 459–69.

Rogerson, C. M. and J. M. Rogerson (1999) 'Industrial Change in a Developing Metropolis: The Witwatersrand 1980–1994', *Geoforum* 30: 85–99.

Rogerson, C. M. (2000) 'Local Economic Development in an Era of globalisation: The Case of South African Cities', *Tijdschrift voor Economische en Sociale Geografie* 91: 397–411.

Rogerson, C. M. (2001) 'Knowledge-Based or Smart Regions in South Africa', *South African Geographical Journal* 83: 34–47.

Rogerson, C. M. (2002) 'Pro-Poor Interventions for Local Economic Development: The Case for Sectoral Targeting', unpublished manuscript, University of the Witwatersrand, Johannesburg.

Rydell, R. W. (1993) *World of Fairs: The Century-of-Progress Exhibitions*, Chicago, Ill.: University of Chicago Press.

Santomasso, E. A. (1980) 'The Design of Reason: Architecture and Planning at the 1939/40 New York World's Fair', in H. A. Harrison (ed.) *Dawn of a New Day: The New York World's Fair, 1939/40*, New York: New York University Press and the Queens Museum, pp. 29–41.

Santos, M. (1979) *The Shared Space: The Two Circuits of the Urban Economy in Underdeveloped Countries*, London: Methuen.

Sardar, Z. (2000) *The Consumption of Kuala Lumpur*, London: Reaktion.

Sassen, S. (1991) *The Global City: New York, London, Tokyo*, Princeton, NJ: Princeton University Press.

Sassen, S. (1994) *Cities in a World Economy*, Thousand Oaks, Calif: Pine Forge Press.

Sassen, S. (2001a) *The Global City: New York, London, Tokyo*, 2nd edn, Princeton, NJ: Princeton University Press.

Sassen, S. (2001b) 'Global Cities and Global City-Regions: A Comparison', in A. Scott (ed.) *Global City-Regions: Trends, Theory, Policy*, Oxford: Oxford University Press, pp. 78–95.

Sassen, S. (ed.) (2002) *Global Networks, Linked Cities*, London: Routledge.

Savage, M. (1995) 'Walter Benjamin's Urban Thought: A Critical Analysis', *Environment and Planning D: Society and Space* 13: 201–16.

Savitch, H. and R. Vogel (2004) 'Suburbs Without a City: Power and City-County Consolidation', *Urban Affairs Review* 39: 758–90.

Schelling, V. (ed.) (2000) *Through the Kaleidoscope: The Experience of Modernity in Latin America*, London: Verso.

Schiffer, S. R. (2002) São Paulo: Articulating a Cross-Border Region', in S. Sassen (ed.) *Global Networks, Linked Cities*, London: Routledge, pp. 209–36.

Schmitz, H. (2000) 'Does Local Co-operation Matter? Evidence from Industrial Clusters in South Asia and Latin America', *Oxford Development Studies*, 28: 323–36.

Schmitz, H. and K. Nadvi (1999) 'Clustering and Industrialization: Introduction', *World Development* 27: 1503–14.

Schumaker, L. (2001) *Africanizing Anthropology: Fieldwork, Networks and the Making of Cultural Knowledge in Central Africa*, Durham, NC and London: Duke University Press.

Schwartz, R. (1992) *Misplaced Ideas: Essays on Brazilian Culture*, London: Verso.

Scott, A. J. (ed.) (2001) *Global City-Regions: Trends, Theory, Policy*, Oxford: Oxford University Press.

Scott, A. J. and M. Storper (2003) 'Regions, Globalization, Development', *Regional Studies* 37: 579–93.

Scott, A. J., J. Agnew, E. J. Soja and M. Storper (2001) 'Global City-Regions', in A. J. Scott (ed.) *Global City-Regions: Trends, Theory, Policy*, Oxford: Oxford University Press, pp. 11–30.

Sennett, R. (1990) *The Conscience of the Eye*, London: Faber & Faber.

Sevcenko, N. (2000) 'Peregrinations, Visions and the City: From Canudos to Brasília, the Backlands become the City and the City becomes the Backlands', in V. Schelling (ed.) *Through the Kaleidoscope: The Experience of Modernity in Latin America*, London: Verso, pp. 75–107.

Shatkin, G. (1998) ' "Fourth World" Cities in the Global Economy: The Case of Phnom Penh', *International Journal of Urban and Regional Research* 22: 378–93.

Shaw, T. (1976) 'Zambia: Dependence and Underdevelopment', *Canadian Journal of African Studies*, 10: 3–22.

Short, J. R., Y. Kim, M. Kuus and H. Wells (1996) 'The Dirty Little Secret of World Cities Research: Data Problems in Comparative Analysis', *International Journal of Urban and Regional Research*, 20: 697–715.

Simmel, G. (1971) 'The Metropolis and Mental Life', in D. Levine (ed.) *Georg Simmel: On Individuality and Social Forms*, Chicago, Ill.: University of Chicago Press, pp. 324–339.

Simmel, G. (1997) 'The Metropolis and Mental Life', in D. Frisby and M. Featherstone (eds) *Simmel on Culture*, London: Sage, pp. 174–85.

Simmie, J. (2004) 'Innovation Clusters and Competitive Cities in the UK and Europe', in M. Boddy and M. Parkinson (eds) *City Matters: Competitiveness, Cohesion and Urban Governance*, Bristol: Policy Press, pp. 171–98.

Simon, D. (1989) 'Colonial Cities, Post-colonial Africa and the World Economy: A Reinterpretation', *International Journal of Urban and Regional Research*, 13: 68–91.

Simon, D. (1995) 'The World City Hypothesis: Reflections from the Periphery', in P. Knox and P. Taylor (eds) *World Cities in a World-System*, London: Routledge, pp. 132–55.

Simone, A. (1998) 'Globalization and the Identity of African Urban Processes', in H. Judin and I. Vladislavic (eds) *Blank__: Architecture, Apartheid and After*, Amsterdam: NAI, D8.

Simone, A. (2001) 'Straddling the Divides: Remaking Associational Life in the Informal African City', *International Journal of Urban and Regional Research* 25: 102–17.

Simone, A. (2004) *For the City Yet to Come: Changing African Life in Four Cities*, Durham, NC and London: Duke University Press.

Sites, W. (2003) *Remaking New York: Primitive Globalisation and the Politics of Urban Community*, London and Minneapolis, Minn.: University of Minnesota Press.

Sjoberg, G. (1959) 'Comparative Urban Sociology', in R. K. Merton, L. Broom and L. S. Cottrell (eds) *Sociology Today: Problems and Prospects*, New York: Basic Books, pp. 334–59.

Smith. D and M. Timberlake (1995) 'Cities in Global Matricies: Toward Mapping the World-System's City System', in P. Knox and P. Taylor (eds) *World Cities in a World System*, Cambridge: Cambridge University Press, pp. 79–97.

Smith, M. P. (1998) 'The Global City: Whose Social Construct Is It Anyway? A Comment on White', *Urban Affairs Review* 33: 482–8.

Smith, M. P. (2001) *Transnational Urbanism*, Oxford: Blackwell.

Smith, N. (2002) 'New Globalism, New Urbanism: Gentrification as Global Urban Strategy', *Antipode*, 34 (3): 434–57.

South African Cities Network (2004) *State of the Cities Report*. Cape Town: South African Cities Network.

Southall, A. (1973) 'Introduction', in A. Southall (ed.) *Urban Anthropology: Cross-Cultural Studies of Urbanization*, New York: Oxford University Press, pp. 3–14.

Stanley, B. (2001) ' "Going Global" and Wannabe World Cities: (Re)Conceptualising Regionalism in the Middle East', *GAWC Research Bulletin* 45, available on-line at <http://www.lboro.ac.uk/gawc>.

Stepan, N. L. (2001) *Picturing Tropical Nature*, London: Reaktion.

Stoker, G and K. Mossberger (1994) 'Urban Regime Theory in Comparative Perspective', *Environment and Planning C: Government and Policy* 12: 195–212.

Stoler, A. L. (1995) *Race and the Education of Desire*, Durham, NH and London: Duke University Press.

Storper, M. (1995) 'Territorial Development in the Global Learning Economy: The Challenge to Developing Countries', *Review of International Political Economy* 2: 394–424.

Storper, M. (1997) *The Regional World*, London and New York: Guilford.

Storper, M. (1998) 'The Limits to Globalization: Technology Districts and International Trade', *Economic Geography*, 74: 60–93.

Storper, M. and A. Venables (2004) 'Buzz: Face-to-Face Contact and the Urban Economy', paper presented to 'The Resurgent City' conference, LSE, London, April.

Stren, R. (2001) 'Local Governance and Social Diversity in the Developing World: New Challenges for Globalising City-Regions', in A. J. Scott (ed.) *Global City-Regions: Trends, Theory, Policy*, Oxford: Oxford University Press, pp. 193–213.

Szondi, P. (1988) 'Walter Benjamin's City Portraits', in G. Smith (ed.) *On Walter Benjamin: Critical Essays and Recollections*, Cambridge, Mass.: MIT Press.

Taylor, P. (1997) 'Hierarchical Tendencies amongst World Cities: A Research Proposal', *Cities* 14: 323–32.

Taylor, P. (2000) 'World Cities and Territorial States under Conditions of Contemporary Globalisation', *Political Geography* 19: 5–32.

Taylor, P. (2001) 'West Asian/North African Cities in the World City Network: A Global Analysis of Dependence, Integration and Autonomy', *GAWC Research Bulletin* 58, available on-line at <http://www.lboro.ac.uk/gawc>.

Taylor, P. (2004) *World City Network: A Global Urban Analysis*, London: Routledge.

Taylor, P., D. Walker, G. Catalano and M. Hoyler (2001) 'Diversity and Power in the World City Network', *GAWC Research Bulletin* 56, available on-line at <http://www.lboro.ac.uk/gawc>.

Taylor, W. R. (1992) *In Pursuit of Gotham: Culture and Commerce in New York*, Oxford: Oxford University Press.

Telford, A. (1969) *Johannesburg: Some Sketches of the Golden Metropolis*, Cape Town: Books of Africa.

Thomas, W. I. and F. Zaneicki (1927) *The Polish Peasant in Europe and America, Vol. II*, New York: Alfred A. Knopf.

Thompson, E. C. (2002) 'Migrant Subjectivities and Narratives of the *Kampung* in Malaysia', *Sojourn*, 17: 52–75.

Thrift, N. (2000) ' "Not a Straight Line but a Curve", or, Cities are not Mirrors of Modernity', in D. Bell and A. Haddour (eds) *City Visions*, London: Longman, pp. 233–63.

Tomlinson, R. (1999) 'From Exclusion to Inclusion: Rethinking Johannesburg's Central City', *Environment and Planning A* 31: 1655–78.

Tsing, A. (1993) *In the Realm of the Diamond Queen*, Princeton, NJ: Princeton University Press.

Tsing, A. (2000) 'The Global Situation', *Cultural Anthropology* 15 (3): 327–60.

Turok, I., N. Bailey, R. Atkinson, G. Bramley, I. Docherty, K. Gibbs, R. Goodlad, A. Hastings, K. Kintrea, K. Kirk, J. Leibovitz, B. Lever, J. Morgan and R. Paddison (2004) 'Sources of City Prosperity and Cohesion: The Case of Glasgow and Edinburgh', in M. Boddy and M. Parkinson (eds) *City Matters: Competitiveness, Cohesion and Urban Governance*, Bristol: Policy Press, pp. 13–32.

Tyner, J. A. (2000) 'Global Cities and Circuits of Global Labour: The Case of Manila, Philippines', *Professional Geographer* 52: 61–74.

UNCHS (United Nations Centre for Human Settlements) (2000) 'The State of the World's Cities: 1999, Addendum to the Report of the Executive Director', available on-line at <http://www.unchs.org>.

UNCHS (United Nations Centre for Human Settlements) (2001) *Cities in a Globalising World: Global Report on Human Settlements 2001*, Earthscan and UNCHS: London.

United Nations Department of Economic and Social Affairs, Population Division, (2004) *World Urbanization Prospects: The 2003 Revision*, New York: United Nations.

Varsanyi, M. (2000) 'Global Cities from the Ground Up: A Response to Peter Taylor', *Political Geography* 19: 33–8.

Ward, D. and O. Zunz (eds) (1992) *The Landscape of Modernity*, Baltimore, Md. and London: Johns Hopkins University Press.

Ward, P. (1993) 'The Latin American Inner City: Differences of Degree or of Kind?', *Environment and Planning A* 25: 1131–60.

Watts, M. (2003) 'Alternative Modern: Development as Cultural Geography', in K. Anderson, M. Domosh, S. Pile and N. Thrift (eds) *Handbook of Cultural Geography*, London: Sage, pp. 433–54.

Werbner, R. P. (1984) 'The Manchester School in South-Central Africa', *Annual Review of Anthropology* 13: 157–85.

Wheatley, P. (1967) *City as Symbol*, Inaugural Lecture, University College London, London: H.K. Lewis.

White, J. (1998) 'Old Wine, Cracked Bottle? Tokyo, Paris and the Global City Hypothesis', *Urban Affairs Review* 33: 451–77.

Willis, C. (1992) 'Form Follows Finance: The Empire State Building', in D. Ward and O. Zunz (eds) *The Landscape of Modernity*, Baltimore, Md. and London: Johns Hopkins University Press, pp. 160–87.

Wilson, E. (1992) 'The Invisible Flâneur', *New Left Review*, 191: 90–110.

Wirth, L. (1964) *Louis Wirth: On Cities and Social Life*, Chicago, Ill.: University of Chicago Press.

Wolfensohn, J. D. (1999) 'Foreword from the World Bank', in *Business Briefing: World Urban Economic Development*, Official Briefing for World Competitive Cities Congress, Washington, DC: World Bank, pp. 12–13.

Wolfensohn, J. D. (2001) 'The World Bank and Global City-Regions', in A. J. Scott (ed.) *Global City-Regions: Trends, Theory, Policy*, Oxford: Oxford University Press, pp. 44–9.

Wolff, J. (1985) 'The Invisible Flâneuse: Women and the Literature of Modernity', *Theory, Culture and Society* 2: 37–46.

World Bank (1991) *Urban Policy and Economic Development: An Agenda for the 1990s: A World Bank Policy Paper*, Washington, DC: World Bank.

World Bank (2000) *Cities in Transition: World Bank Urban and Local Government Strategy*, Washington, DC: World Bank.

Yeoh, S. G. (2001) 'Creolized Utopias: Squatter Colonies and the Post-Colonial City in Malaysia', *Sojourn*, 16: 102–24.

Young, M. and P. Willmott (1986) [1957] *Family and Kinship in East London*. Harmondsworth: Penguin.

Young, R. (1988) *Zambia. Adjusting to Poverty*, Ottawa: North-South Institute.

Yúdice, G. (1999) 'Rethinking the Theory of the Avant-Garde from the Periphery', in A. L. Geist, and J. B. Morleón (eds) *Modernism and its Margins: Reinscribing Cultural Modernity from Spain and Latin America*, London: Garland, pp. 52–80.

Index